AF375219

Mise en Mode

Published 2026 by Design Systems House, LLC, New York, USA.
All rights reserved.
ISBN: 979-8-9938672-0-5

Cover Design: Jennifer D'Amato
Foreword: Luis Ouriach
Editor: Joseph Staples
Illustrator: Erik Loño
Typefaces: Young Serif by Bastien Sozeau, Libre Franklin by Pablo Impallari, and DM Mono by Colophon Foundry.

Mise en Mode was written by Donnie D'Amato and reviewed by Luis Ouriach, ToniAnn Drenckhahn, Frank Stallone, and Katie Langerman.

Please send errors to: errata@ds.house.

Table of Contents

About the Author

Donnie D'Amato is a design systems architect with over 25 years of experience making things for the web. He has led efforts for several companies over his professional history, both as an embedded team member and as a strategic consultant under his company Design Systems House.

He challenges the belief that all design systems should have permission to be functionally different. He aims to discover ways of customizing experiences through a delicate balance of meeting user expectations while also subverting traditional education. He shares his insights with the design systems community through articles, projects, and events.

Acknowledgements

A large thank you to Joe Schmitt as the originator of intents. His initial insight into a system of naming design tokens that separates the value from the purpose has been instrumental in my overall thinking in the design systems practice. From the path he initially laid, I have taken this idea to understand what the concept of semantic truly means to support further flexibility in our expressions.

I'd like to thank Sean Rice as a close partner and sounding board in the many thought experiments that led to Mise en Mode. He is a systems-thinker looking to empower folks who want to express themselves. His open-mindedness helped shape what could be possible in organizations that require many expressions for a variety of reasons.

I'm thankful to Jina Anne for not only coining the concept of design tokens but for also giving me the opportunity to present Mise en Mode to an audience of industry professionals at Clarity 2023. This was my first large-scale conference talk and the response that came from it was overwhelming. So much so that the response was the largest motivator in the creation of this book. The interest that grew from this exposure validated that more folks would benefit from learning about this approach.

I want to thank Ethan Marcotte for our initial conversations about book publishing and for sharing many of his resources that have helped in his process. Moreover, I'd also like to thank him for being an initial inspiration about how I could turn creating on the web from a hobby to a serious career. The elegance of having one page that could respond to a device size was a paradigm-shifting design solution, and I hope Mise en Mode can hold a similar status in the future.

I'd like to thank my trusted readers: ToniAnn Drenckhahn, Frank Stallone, Katie Langerman, and Luis Ouriach. I understand how challenging it is to juggle a work/life balance while volunteering your time to this project. I greatly appreciate that time and the feedback that helped shape the content of this book for future readers. A special thank you to Luis for his additional contribution to the foreword.

I'd like to thank Joseph Staples and Erik Loño for their expertise in editing and illustrations, respectively. These were the roles in completing the book that I knew I couldn't execute myself with high quality, so I am very thankful for being introduced to them by way of the larger tech community. 10/10 would hire again, no notes.

Speaking of community, I'd like to thank everyone that I've interacted with inside of the design and engineering communities. These are the discussions that help us all land on the best approaches to our problems. I'm confident a significant percentage of these interactions have helped direct the thoughts behind Mise en Mode one way or another.

I'd like to thank the folks at RYCO's Escape Room & Lounge. The majority of this book was written there as an escape from the distractions at home. Thank you all for the warm hospitality, free Wi-Fi, and a space to sit and focus on this work.

A big thank you to my wife Jen. Her support through all of my focused research, after-hours discussions, and time away from home to write demonstrates the loving partnership that I never knew I needed. I'm also grateful for her background in print media and her contributions to the finer details that helped complete this book, such as the cover and typesetting.

Finally, I'd like to thank my parents for the freedom to be a creative child.

Foreword

Quality is a differentiator. The demand and requirement for quality on software makers is a constant that won't change.

No matter what new record-breaking industry wave crashes against the ship of software creation, systems should act as a protective and predictable barrier. Quality in practice manifests as something that can be accurately repeated at scale, across timezones, and should survive uncertainty. The magnitude of these expectations can paralyse teams. Under-resourced and over-stretched, we build what we can with what we have and hope that with all the effort we put into our practice the work is respected and consumed accurately.

Quality in systems is a long and patient game of conscious and careful refinement that delicately balances organisational culture, short-term priorities with long term strategy, rooms of egos, and a tech landscape that can't sit still. What is reassuring is that there is a constant — the values we curate have a profound impact on the entire foundation of the products we are selling. Not only is this exciting, it's profoundly important to get right. And do you know what? We can get it right.

Tokens are our purest form of controlling quality. Without them, we sacrifice the ability to pivot quickly and in a cost-effective manner. With them, we can attempt to produce the repeatable and predictable levels of quality that any ambitious, modern company demands. This is why we find it difficult. With infinitely flexible ways of building token systems, we often end up in scenarios where expert design system managers build hyper complex, yet difficult to comprehend architectures. It's why I believe we need to reflect: is what we have been led to believe up to now the best way to achieve our goal?

Can we work towards a sharper standard, prioritising extension instead of addition? A shared agreement that what we're doing now is not serving those responsible for building interfaces in the best possible way. We bloat our systems with a "just one more token" approach that inevitably requires refactoring and remodelling, as we face the dreaded realisation that adoption isn't where it needs to be to secure further team funding.

Complexity leads to a lack of comprehension, and a dearth of understanding leads to deviation.

I have sat in a luxurious position for the past few years, seeing hundreds of design system teams across the globe, from the simple to the wildly complex. From my bird's-eye view seat of the industry, I have witnessed confusion, dissatisfaction, and deep feelings of burnout. I believe this has stemmed from our lack of a shared standard, forcing systems maintainers into an exhausting uphill struggle where the only shared standard is that of proving why you need to exist.

As the ship rocks back and forth amidst an industry in flux, what is reassuring to me is that systems are absolutely critical to the success of any business that wants to experiment at scale. The more we can fix our core values, the more deterministic we can be with our outputs. The more software that floods the market, the more that brand expression matters.

The more we reverse out of the idea that a system is a siloed item on a to-do list, and start placing it at the beating heart of a company's ability to succeed in an unpredictable climate, we realise that systems are brands and brands are systems.

What we have the opportunity to do right now is grab a hold of the wheel and do what we've always wanted to do — steer the ship, putting the system at the centre of every generative output that our hearts desire. Systems, and quality, are everyone's business.

This is why the world is ready for Mise en Mode. What we require is a loosening of the grip on the idea that systems are the bottleneck, because when we look in the rear view mirror of the system, we will see reflecting back the need of an entire business; desperate to work more efficiently with the brand.

We have moved beyond software as the only recipient of our tokens and components, because the ask now is holistic flexibility. That may sound scary, and it certainly can be, but flexibility can still be constrained. Donnie's approach allows us as maintainers to set a wide scope on what's possible, and our consuming colleagues to feel the control and freedom that a system should promote.

Enjoy the ride, the fresh ocean air awaits.

— Luis Ouriach

Preface

From a young age, I was interested in designing experiences that respond to an audience. From building moving models with simple plastic bricks to constructing homecoming floats and winning regional robotics competitions, making things come alive always seemed magical. While experiencing dynamic works of art, a person's face can shift between several emotions, and they don't need to be an expert to enjoy the craft. Creating experiences that are moved by people, figuratively and literally, brought me fulfillment as a designer. With the introduction of the Web, it became much easier to share and iterate expressive work for the world to see. Years ago, creating within this new digital medium was uncharted territory, but the benefits were clear. Crafting user experiences was now less expensive and more sharable than ever before. I'm lucky that this love has turned into a prosperous career.

Fast forward to today; I've been fully enveloped in the practice of design systems, an exercise in taking my expertise in experiential design and sharing ideas with a vibrant community of professionals. We tackle challenges that impact millions of people daily through our decisions, often breaking new ground to answer questions never before asked.

The company I founded, Design Systems House, is dedicated to the future of design systems. We work to find answers to these problems through research, discovery, outreach, and prototyping, among other activities. This book will describe one of these journeys at length, ending with an approach that makes your job easier and empowers your peers to be more involved in design systems.

One of the most important motivations behind my career is to discover solutions within the design system practice that are meant to work for everyone. The Internet is full of recommendations, but they often target particular teams and the individual problems they have identified. This lens has caused our community to believe that design systems are largely meant to be different and only in support of the organization they are developed within. It is now exceptionally challenging for new teams to grow their practices. Putting myself in their shoes, I have to ask: How could someone new know where to start if every design system differs? Shouldn't a system meet standards and expectations universally?

This book aims to provide guidance for both those starting in design systems work and for those experienced who want to reconsider their traditional education. It will revisit some of our design foundations to help inform the unique techniques suggested to support our peers. Eventually, it will describe how we might support our peers' freedom of expression while maintaining our system of thoughtful guidelines. All the while keeping the peace between strong opinions.

The approach that you will eventually discover in this book is designed to support large organizations or product platforms. The complexity that can come from these entities requires considerable resources to have a successful outcome. For smaller sites or teams, what we'll discuss may be overkill as a solution. However, the concepts introduced remain important no matter a company's composition and can be used to inform less critical decisions or prepare for significant growth.

Much of the book has been written through a hypothetical lens; a perfect world where your peers align to the same concepts across the organization uniformly. In practice, people are messy.

While it might be easy to discount these perfect scenarios as unrealistic, using them as a target to work toward should prove beneficial over time. Who knows, you might even achieve the fabled perfect organization alignment.

This book assumes that you are familiar with modern digital design practices, especially the concept of design tokens. In the chapter dedicated to this topic, we will fully realize their purpose more deeply than most have had time or interest to consider. This serious understanding will be critical to fully grasp the final recommendations provided by the end of the book.

It is not uncommon for tools to move at a slower pace than ideas. So, in an attempt to keep this book evergreen, it will avoid specific products or tools and instead demonstrate through more abstract examples. The concepts are paramount, and we can only hope our tools can catch up to support them.

In areas where code is used to demonstrate the execution of an idea, we will be leveraging the syntax of web standards as of the date of publication. This is where my personal expertise lies. Other platforms and technologies could most likely develop their specific methods of execution using the foundations suggested within this book, but they will not be mentioned otherwise.

With that all out of the way, let's begin our journey into *Mise en Mode*.

## PART 1

Concept

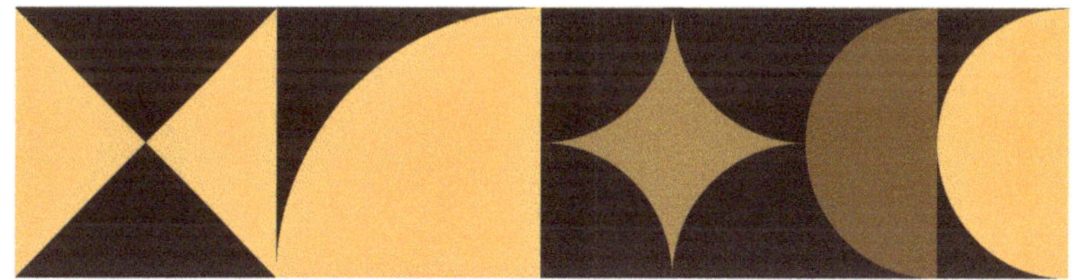

Freedom of Expression

Freedom of expression and design systems are often thought to be on opposing sides. One person wants to explore novel ways of presenting an idea, while another cements well-known practices that have proven successful for users. How can a designer spread their colorful wings when locked in a cage of historic guidance? This is where we see a poor understanding of design systems from outside of our practice. We look like police, out to reprimand every button using the wrong color while designers smuggle purple gradients into their artboard's dark corners.

Let's set the record straight right now. Ask any design system maintainer for their opinion about the color of your buttons. You should find that their response involves accessibility and, more importantly for their job role, agreement. To codify agreement across an organization. Getting people—especially creative people—to agree is very difficult. Creative people often have feelings that manifest in their work. This can manifest as a lack of well-communicated motivation past these emotions. They may say, "This blue looks better" but without understanding *why* it looks better. Design system maintainers have no stake in this game and would rather sell tickets to the "Battle of the Blue Buttons" to crown a winner as long as we can read what the button ultimately says. Stakeholders outside the system should greatly influence the final presentation of the experience. Experience designers focus on the flow from one context to the next, while user interface designers curate the style of known patterns.

As design systems maintainers, our challenge is to support freedom of shared expression while governing an agreed-upon cohesive presentation. How might we allow our peers to experiment with new stylistic mediums while battling with different, but valid, opinions?

If what I've described is too abstract, let me paint a picture of what we'll set out to solve.

Imagine we have a page expected to show pricing options for our product. We offer three tiers on this page: Free, Pro, and Enterprise. Each tier displays a list of features available after purchase, all with a similar layout using brand colors and typography to convey to users that this is a legitimate business they can trust.

Pricing page with 3 tiers of increasing visual treatment

Creatives are taught to curate a more interesting design that attracts attention to the expensive tiers. The psychology behind this is that a person doesn't want the boring tier; they want the vibrant one because it looks like more effort was put into that tier, including its features. We bet on this user behavior and hope more people gravitate to the expensive tier for purchase. This is a common approach to pricing pages and can be seen all over the Web.

For many design system maintainers, supporting this varying visual treatment is a nightmare. At first glance, this looks like it requires design tokens specific to the pricing page and more specific to the tiers. We'll need tokens that describe the new background, foreground, and collection of newly introduced accents to draw attention in ways that aren't found anywhere else in the product. For some, it just seems easier to mark this as an exception that exists outside of the system because it is traditionally too special of a case to support. This causes other exceptions to be identified, and before you know it, you wonder where the system is even useful anymore. For veteran maintainers, this is several units of work for an exercise in token naming; assigning new values with resources that need to be published, documented, and later maintained, all for a single place in the entire ecosystem. The juice just doesn't seem worth the squeeze.

This example is what has many creatives at odds with their system maintainers. From the maintainer's perspective, it seems like the creative is being *too* creative. "If they were just a bit less creative, they could align better with the rest of the system." We want our users to trust our offering, and the best way to do that is to meet our users' expectations across the experience, every step of the way. Any deviation from the norm is at risk of a user questioning the product's legitimacy, losing confidence, and leaving with one swift click. The maintainer believes the system should support a product's trust by offering repeatable patterns and tested standards.

From the creative perspective, the expression is warranted. This additional treatment is meant to enhance the messaging, with a root in human psychology and marketing best practices. "This is a valid expression to support, so why doesn't the system account for it?" What else does it not account for? If the system can't support this change, how can I trust the rest of the system will support my creativity over time? "The system isn't supporting me!"

Recognize that in this scenario there are conflicting requirements. The business wants more significant sales, while the user wants assurance that this company is legitimate before opening their wallet. Compromises must be made to balance both of these needs, but the sacrifice does not need to be extreme. We should be able to have our design system and express it too.

This has become more apparent recently with the need to support multiple brands with a single system or even offer "dark mode" for the organization. We are seeing changes to the design landscape that have never been considered before in the digital medium, where outside influences can affect the final presentation of our work. It is not enough to design responsively for mobile and tablet, but to also consider translated content, assistive technologies, and other permutations of personalization options. In a striking study by the Common Sense Advisory[1], users are twice as likely to purchase a product when it is presented in their language. With this fact, we should begin to see the benefit of adapting to users in various ways, even if only for profit. What you design is not always necessarily what the user wants. The design needs to be flexible enough to adapt to any number of changes that could occur, either by business need or user preference or anything in between. A well-designed system will be prepared for these possibilities, while a well-experienced creative will be humble to these eventualities.

The big question is, assuming we want designers to join us in the system, is adding more flexibility to support freedom of expression attainable without causing unnecessary complexity or maintenance? I have confidence that the answer is yes, and by the end of this book, you and your team should have a new worldview of design that empowers your organization to be creative in useful ways.

Endnotes

1. *https://csa-research.com/Featured-Content/For-Global-Enterprises/Global-Growth/CRWB-Series*

Design Tokens

The first step in creating a flexible design system meant to support different expressions is to separate the decisions about the way the experience appears from the structure that supports those decisions.

Imagine you move into a new home. Over time your new surroundings will reflect more of your personality through lots of small changes, but the structure of the dwelling hosting your style should change very little; you can add new curtains but can't move the windows. OK, you could move the windows, but not without a great deal of effort. You have to weigh your options between purchasing some cheap window dressing or restructuring the walls. We do the same when determining how to customize our experiences.

Using the above analogy, **design tokens** can be thought of as the way we define what parts of our house are going to receive personality; things that can be changed with very little resource and potentially make a big impact. Painting the walls a very dark color makes a statement, and so does making your interface dark in color. Jina Anne, who coined the term "design token", defined them as "design decisions." Importantly, there's more than one decision in a token. There is a decision for where personality should be customizable and then a decision for how it should be customized. This is like developing a house made of titanium and needing to place the curtain rods before the first person moves in. Decisions like these will be costly to change later. You are deciding the placement of the rods, but the person living there will decide what curtains will go in those places.


```
$color_teal_500
```


While an easy analog to the concept would be a programming language variable, the key part of the given definition is the word "design." This makes the intention of the design token to be meant as a design artifact. The difference could be considered minimal, but I believe this restricts the use of design tokens to describe the design as data specifically. That decision is meant to be shared across the digital platform as a variable and, more importantly, as a shared agreement across teams.

In other words, we as a team agree that a token named `$color_teal_500` is our medium-shade teal and we can all use this token to represent this agreement in our daily work.

If you've spent any time within design systems, either using or maintaining them, you may have come across different types of design tokens. These types are commonly discussed as "tiers" because the kind of token is meant to sit at a particular spot in a larger ecosystem of token architecture among other similar tokens. There are typically three tiers that you may come across when researching design tokens.

Primitive Tokens

Primitive tokens are the design decisions closest to the value they represent. The word "primitive" comes from the term's use in programming, where "primitive values" are foundational building blocks for a programming language meant to have a permanent meaning in their ecosystem. In design tokens, the primitive token for color would have a hex color value directly assigned. As an example: `$color_teal_500` could be assigned a hex value of `#11FFCC`. There are two main reasons why these tokens exist.

Hick's Law[1] says "the time it takes to make a decision increases with the number and complexity of choices." There are thousands of possibilities when it comes to choosing a color of teal. Reducing the number of colors to choose from makes it easier to choose. Primitive tokens typically provide a group or scale of tokens all within the same family with slight variations. People choosing which tokens to use gain more confidence by picking from a predetermined set that should have been curated by someone who has more experience choosing color from the thousands of possibilities.

Additionally, it is much easier to communicate these values as human-utterable token names than their cumbersome coded values. It is much harder to say "hash-one-one-eff-eff-see-see" than it is to say "color-teal-five-hundred." Simplifying these names provides more familiar meaning of the color value to a human, while maintaining the value that a computer will eventually require. If you are working with a team of people that need to choose colors regularly, creating nicknames for those choices that are understood within the team can be helpful.

You could imagine that if a new value was added to an existing nickname that it would cause quite a stir. In the case of design tokens, `$color_teal_500` should certainly always be a teal color, and we'll hope that it is a teal somewhere in the middle of the spectrum. At first, it could seem reasonable to add this as the background color to all of our primary buttons. However, let's consider what happens during a rebrand where this teal is no longer our button color; it's now orange. We can't simply change the value `$color_teal_500` to orange because that breaks our expectations for what the name represents. It would be like asking for the double-walled blackout curtain and getting a bath towel. In order for us to update all of the places where `$color_teal_500` is used to be orange, we'd have to find all of the places one-by-one and manually update them to be orange. This defeats the purpose of using the concept of a variable in the first place. You might as well be using hex codes like it's 1999.

Maintaining the association of the name to the value continues to be important across light and dark modes. It is difficult to hold the context of `$color_teal_700` being a light teal in one context and a dark teal in another. In this way, each time we communicate `$color_teal_700` we would need to include further context about which teal it represents. At this point, it would be more clear to include that context in the name: `$color_teal_700_dark`. This becomes confusing since "dark" could be referring to the color or the context where the color is used. This highlights a defining feature of primitive tokens: they are not context-aware. `$color_teal_700` should always be the same value no matter what later influence may come. Once you set a primitive token to a value, it should be nearly permanent and clearly communicate the value it represents no matter what context it is used in. Later token tiers can be influenced by context and select a primitive to use.

The primitive token tier is at the bottom of our tree of tokens because the value set here will propagate up to provide a value to other higher tiers. It is the tier that is farthest from the components found in a user-interface because a change here will cascade across the platform at large, and we want to be sure that this does not occur without understanding the usage and consequences. As mentioned, changes at this tier shouldn't occur, but additions or substitutions could be appropriate based on how frequently design decisions are made.

Component Tokens

At the top of the token tree and at the tips of its branches are **component tokens**, which describe specific stylistic parts of the components that could have a variable design decision. These are called "component" tokens because they are assigned to a single specific component with its value informed by more generic tokens. An example could be `$modal_foregroundColor`, which expects a color value for the text found within the modal component and *only* the modal component. Here, we could assign a token as a value such as `$color_black` or even `#000` directly. This tier of tokens is fairly easy to identify as the set of tokens for any single component within your library, unique to its individual existence and style.

The reason that these tokens tend to exist is to support highly customized parts of components. If there's a need to curate the modal's text differently from most other text across the interface, then the component tokens seem to be the logical choice.

These tokens should be avoided for a few reasons. The first reason is to reduce bloat. Imagine a single component and all of the component tokens that would be needed to cover the variable style attributes that this one component needs. Background colors, font sizes, corner rounding, etc.—all for one component. If you multiply this number by the number of components in your library, you will get a rough amount of component tokens you'd be responsible for maintaining. This number can easily be in the thousands across many substantial component libraries.

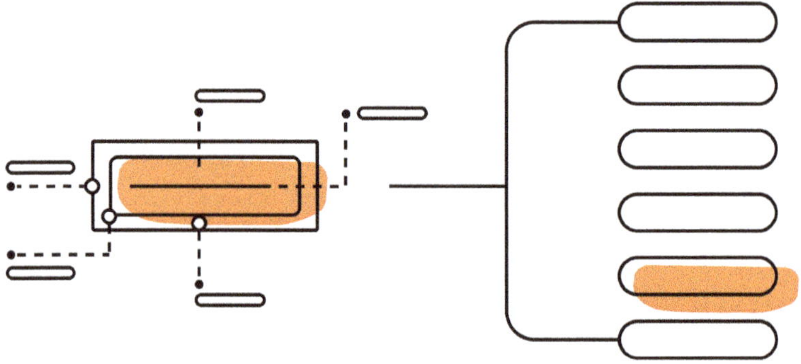

The anatomy of a button's style properties multiplied by the components in the library

Another reason to avoid component tokens is to maintain consistency. The more component tokens that exist, the more ways the experience can deviate from a consistent presentation. Imagine every component receiving a slightly different foreground color as opposed to having a limited assignment of foreground colors applied more generically across the experience.

If every component had a unique token for every design decision that could be controlled, you can easily have thousands of tokens that need to be curated across your design system. If you were making these decisions individually, this would make maintaining a component token ecosystem untenable. A better approach would be to summarize this set into more generic categories to lower the number of tokens introduced while keeping a consistent treatment across similar components. Component tokens can be introduced when describing something uniquely specific; not appropriately covered by a more generic token. These should be considered outliers, and we'll explore legitimate examples later.

The Benefit of Tokens

Before we explore the middle tier, I'd like to address a misconception.

As mentioned earlier, an easy mental model to align design tokens with are variables in computer programming. When writing a program, you can assign a value to a variable. That allows the name to be reused; representing the value in other areas of the ecosystem. While this is a feature of tokens, I don't believe it is the primary benefit.

In computer programming, the person managing the reuse of the variable is an engineer or a team of engineers. An engineer has all the necessary knowledge to reuse that variable across their platform. In other words, the variable doesn't need to exist across roles. Variables are squarely an engineering concept.

This is different for the design token ecosystem, where these are variables that must exist across roles, most often between design and engineering teams. So the benefit isn't necessarily reuse, but to act as sort of a wormhole between the design and engineering galaxies. The value that a designer chooses can affect something in the engineering world. This is very important because of the different responsibilities between these roles. A designer—or more generally a stakeholder in the visual presentation—is responsible for the value, while an engineer is responsible for the capsule which propels the color into the final digital product.

Tokens remove the bottleneck of asking an engineer to update a color that isn't their responsibility to choose themselves. Tokens empower design stakeholders to make visual changes without relying on engineering skills or alternative priorities.

OK, back to our regularly scheduled programming.

Intents

The tier commonly considered as the middle ground between primitives and component tokens is what has been commonly called "semantic" tokens. The name was given to this tier to suggest that these tokens have meaning. If you are familiar with these tiers and the history behind the names, then please continue to keep your current understanding. For folks who might be confused about the word semantic, I will rename this tier for the remainder of this book to help disambiguate the term. It is common for beginners to misunderstand the word semantic here. If you consider the purpose of the word semantic broadly, you'll lose the intention of this specific tier. Every token in any tier could be called semantic because they all mean something.

To avoid tangential discourse about the word semantic, I'll use the word **intents** to describe this tier. Joe Schmitt coined the term intents, and this set of design tokens is meant to convey design intention within the experience. In other words, the expectation of this tier is for a designer to look at an object within the experience and say, "What is this object's intention?" or more simply, "Why is this thing here?"

Thinking about a design token in this way is a powerful paradigm shift because it holds designers accountable for design decisions. If a designer cannot describe the user-centered intent behind an object, then it is a sign that this object isn't important to the user experience. It separates user needs from egocentric aesthetics.

Intents are different from the other tiers because they are meant to describe user experience needs and not individual objects. For example, you may have an intent category describing all form inputs and their appropriate properties. As a result, you may consider `$control_borderColor` to be assigned to the border color of checkboxes, radio buttons, text fields, and select menus. We are assigning a border to describe the general concept of a user entering information into the system.

This is beneficial for a few reasons. First and foremost, assigning the same style value to all components of a particular intent category conveys to the user that, when given a similar treatment, all of these components are related in some way. This makes it easier for someone using your interface to identify how they should progress and where to look next. This consistency builds trust between the user and the experience, letting a user know that when the next input is introduced, it will look and, more importantly, behave like all other inputs they've come across before. When components of a particular category don't seem to align, a user could question if the component they are interfacing with is truly a part of this larger experience which may cause them to reconsider moving forward at all. Any time something subverts a user's expectations within the experience, it is a chance for them to eject themselves from it. Our goal in user experience design is for the user's journey to be seamless without unintended friction that can occur through deviations in design.

The other reason is curation and reducing maintenance. As we've covered earlier, the more tokens introduced, the more values must be curated and guided in that selection. The goal should be to generalize the tokens to cover multiple components and their similar styles. Because an intent is generalizing a category of component family behavior and purpose, it is meant to cover the style property for more than one component. Having an intent assigned to multiple components reduces the number of tokens needed to describe the component's styles.

If primitive tokens are only needed to communicate values, and component tokens should be avoided, this means that intents are the most important tier of tokens to manage.

We'll be focusing on intents for the remainder of the book and in places where I use the word (design) token from now on, I'm

referring to an intent as we've now introduced them unless
mentioned otherwise.

Constructing Intents

Folks who have seen design tokens before may have some
familiarity with variations of token naming patterns.
Organizations have included parts like domain, category, state,
and variant, among other possibilities to help describe an
individual token. The construction of an intent is very simple,
having only three parts in most cases: purpose, priority,
and property.

Purpose

The exercise of identifying purpose is meant to target the
generic existence of the visual element. For example, in a certain
experience, if you need your user to include a mixed amount of
text in short form, providing them with only a single button
would be inappropriate. The button would not accurately match
the expected experience of freely typing. There is a better
pattern to provide the user, such as a text field.

But the question you need to ask yourself here is, "Why is the
text field the most appropriate pattern for this?" It might seem
obvious, but the exercise of understanding *why* a pattern is
more appropriate helps to uncover what the intents could be.

An example answer could be, "I need the user to provide their
address," and the system could suggest a specifically designed
"address input" component. While this is possible, this is
probably too specific. It would be more efficient to use a "text
field" component instead. In other words, considering a less
specific component reduces the number of components the
system needs to maintain, and it reduces the cognitive load of
learning an otherwise newly introduced component. Just by
adding unique messaging for context to the experience, the
same text field can satisfy this unique address field need.

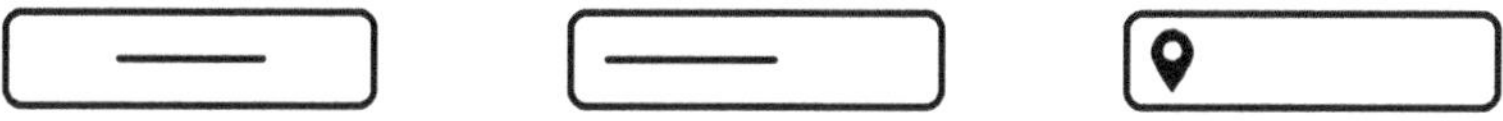

*A button, a generic text field, and an address field
in order of less appropriate to most appropriate*

This exercise of grouping multiple components under a shared purpose is not just beneficial for maintenance and user familiarity. It is also the core principle behind intents. We can describe the component's stylistic properties with a standardized naming convention, supporting the large majority of all user interface concepts. This keeps the maintenance of tokens low, while meeting user expectations within the experience.

For example, the **control** category of intents is probably the most clear to identify. This is meant to cover all form input elements, commonly called controls. These are areas of the interfaces where we expect a user to input information. This could be as simple as a true/false boolean or as complex as a multi-selection pattern. Elements of this family should have similar styleable attributes and be presented with similar values to suggest user input cohesively. In other words, all of the borders of these elements should be given the same treatment to suggest to the user that these elements are related and serve similar purposes, even if the specific action in the context is different. Just because the form has an address field and a postal code field with different expectations of the value the user is meant to input, the treatment of those inputs should be very similar to show they are part of a larger cohesive experience.

The name of the category "control" is meant to generalize a family of similar components. This doesn't necessarily need to begin with a visual similarity but instead focus on an experiential similarity. A user should be able to identify an experiential pattern presented with a purpose using the visual treatment, whatever that will eventually become. The visual similarity comes from a result of the generic family shared across similar components. As an example, the checkbox and the email input both have the same generic purpose: to accept user input. While you could name the category "input", it is important to avoid unintentionally excluding components like the checkbox from that category. Other words like "form" or "field" also suggest individual components which could suggest exclusion of components from the family.

Priority

Most design systems will have a collection of buttons. In that collection, the concept of a "primary" button is most likely to be used. The purpose of the primary button is not to be a container for a primary brand color but instead to be the primary action that a user should be taking in their current experience. This is why it is commonly recommended only to have a single primary "call-to-action" button on any page. This is the action we either want or expect the user to take. If multiple call-to-action buttons existed, each with a different goal, the user would be confused about what action to take and perhaps go in a direction they did not intend to go.

So if the primary button is the primary action to take, and there should be only one on the page, the "primary" part of the name is meant to represent priority. As in, this button treatment indicates the most important button. The button is then styled to stand out from other buttons, so it is clear that this is the most important element to pay attention to at this moment. This could be the "compose" button in an example application or an "unread" button in a messaging platform. Research and data should drive which button on any interface should be the primary action, and all other buttons should have less priority.

I commonly call this category **action**, as it is an interactive element where the user can execute a command with a single press, like buttons and links. While I'm squarely in the camp that links should be given a visual treatment that indicates a click will navigate, the reality is that users are becoming accustomed to buttons that navigate. For example, a login button commonly redirects the user to an authenticated experience or a checkout button sends the user to complete a payment. If you feel strongly about separating the concept of execution from navigation, feel free to introduce a "navigate" category with separate intents for distinct visual treatments from "action." However, recognize that these may visually merge over time, and you'll need to support both once introduced. It might also be appropriate to allow the primary and secondary visual treatments to be "button-like," while the default treatment is "link-like," if you want to avoid a new category. The action category includes an additional qualifier for priority in its construction. `$action_primary_backgroundColor` will describe the primary button's background color as different from the `$action_secondary_backgroundColor`. You'll find this construction to be the most common across these tokens: `$purpose_priority_property`.

This is why the "control" category is easiest to identify, as it does not include priority. When a form exists in an experience, it does not generally convey any priority in a similar way to the "action" category and its components. All forms are effectively at the same level of importance within the experience. While one could argue that controls with errors would have more priority, we'll be able to emphasize this in another way without introducing new tokens.

The next category to introduce is the **surface** collection of intents. This is meant to support commonly non-interactive containers of content. This could describe the entire page or a small section of it. Importantly, this category also includes priority. This means that the `$surface_primary_backgroundColor` represents the most important section of the page. This is often conveyed through elevation, presenting the most important content closer to the user in three-dimensional space. This does not need to be restricted to shadows alone, as it is common for a surface to appear brighter the closer it is to the viewer. As you may expect, the foreground color to appear on that surface heavily depends on the background for accessibility purposes, so these decisions must be made in unison.

It is uncommon to think of surfaces in terms of priority verbatim in traditional user-interface design. However, it is not uncommon to find examples of tokens in public design systems that describe the quality of `z-index`. This is effectively a way of conveying priority where the primary surface is most likely used for your disruptive modal, as it should be the most important content when it appears. When thinking in terms of `z-index` for a modal, we'd typically think of a very high number so it appears on top of all other content. Here, we are instead describing the concept of most important, so that is conveyed to the user as the most important element. As another example, your secondary surface is used for flyouts that appear temporarily on top of other content to make a quick decision before returning to the base elevation.

The last priority level is called "auxiliary." This is meant to convey that this level is necessary but not in a hierarchy of importance. You could think of this as subtle or passive but I've found this diminishes the necessity too heavily. Especially since the `$surface_auxiliary_backgroundColor` describes the main page background, which is important to include but not important to the user in terms of focus or achieving a goal. It is simply a necessary surface where all other elements will be placed. You may be tempted to rename this as "base" but I'd argue that doesn't convey a level of importance. We'll also soon explain a benefit of separating the concept of setting a default from the naming of these tokens.

At this point, you may consider these priority levels too restrictive. You may imagine that your experience has more than three levels of priority that need to be accounted for. You may have even seen other systems with more than half a dozen choices for elevation. In response, I'll ask you to reconsider this from the user's perspective. If more than three priority levels are in one interface, how could a user accurately determine what to do next? We want to keep the number of priority options low. This restricts the user to follow the best direction without a battle for attention.

The final common category is **text**, which describes the attributes of text found within the experience. This category, too, includes a priority similar to what has been described in other categories. An easy analogy is heading sizes in HTML. The more important the message, the more it should grab attention visually through some stylistic choices. At first, you may believe that you require many more levels of priority to cover the intention of your text within an experience. However, we will see in a later chapter how to get a flexible typography scale from a limited set of intentional priorities. For now, you can think about how we rarely use `<h4/>` HTML elements because we do not typically need to provide such granularity to our interfaces. Instead, we might consider separating the interface further to allow users to focus on the important content rather than bombard them with additional content hierarchies.

*The hero section is more prominent than the
smaller card underneath in size and placement*

An example of an intent in this category might be `$text_
primary_fontSize`, which would be appropriate for headlines
and titles. What might not be immediately clear is `$text_
secondary_fontSize` as expected for body text. This is because
the last priority `$text_auxiliary_fontSize` is meant for smaller
details commonly meant for legal copy or form field help
messaging. The recommendation works this way because it is
meant to ensure that the provided font metrics to describe the
default text styles are the least prominent option possible while
remaining accessible. In other words, the `$text_auxiliary_
fontSize` must be the smallest accessible size. Any size higher
immediately conveys a higher priority by design. Any size lower
is in danger of being illegible.

This also highlights that you should choose which purpose and
priority should be considered the default for a particular
category representing a group of components. As an example,
the most common treatment for surfaces should be `$surface_
auxiliary_backgroundColor`, the most common treatment for
text is `$text_secondary_fontSize`, and the most common for
actions should be `$action_auxiliary_backgroundColor`. Notice
the priority differences. We identify what's the most commonly

used by what we expect to appear most often in an interface. Don't encode that expectation into the token name. Instead, make this a setting in a component that uses these tokens.

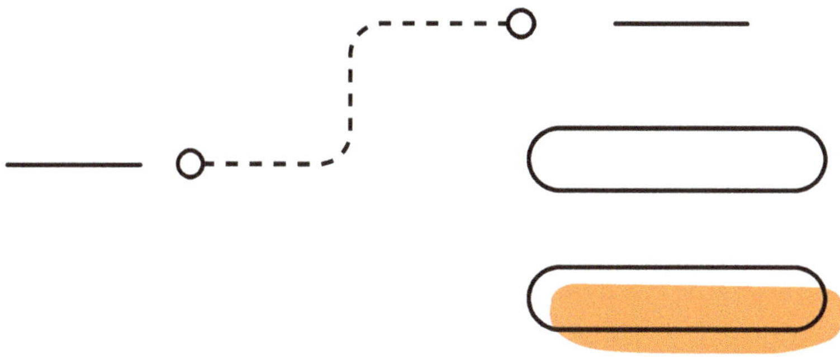

*Set what default means at the
component, not inside your tokens*

The "text" category is somewhat unique, consisting of separate properties from the other categories we've introduced. This category will include font size, font family, font weight, line height, and other metrics related to typography. Color is not recommended here, as the expected color is more related to a container's background color than the font styles. That color could also be appropriate for iconography or other glyphs, which are usually unaffected by font metrics.

Property

We've already seen some examples of properties, specifically for color and typography. Affecting these properties typically provides a great impact when attempting to convey some new expression. Flipping the black to white and vice-versa would produce a stark contrast for an interface, literally and figuratively. Any of these "purpose plus priority" intents could receive any additional properties to further style the experience. However, some of these will not fare well for the system. We'll review additional considerations in later chapters. For now, we have a good set to start from.

The Benefit of Intents

If we were to enumerate what we've considered, so far we would
have the following:

Form controls (3)

```
$control_backgroundColor
$control_foregroundColor
$control_borderColor
```

Buttons and links (9)

```
$action_primary_backgroundColor
$action_primary_foregroundColor
$action_primary_borderColor
$action_secondary_backgroundColor
$action_secondary_foregroundColor
$action_secondary_borderColor
$action_auxiliary_backgroundColor
$action_auxiliary_foregroundColor
$action_auxiliary_borderColor
```

Non-interactive containers (9)

```
$surface_primary_backgroundColor
$surface_primary_foregroundColor
$surface_primary_borderColor
$surface_secondary_backgroundColor
$surface_secondary_foregroundColor
$surface_secondary_borderColor
$surface_auxiliary_backgroundColor
$surface_auxiliary_foregroundColor
$surface_auxiliary_borderColor
```

Font metrics (12)

```
$text_primary_fontSize
$text_primary_fontWeight
$text_primary_fontFamily
$text_primary_lineHeight
$text_secondary_fontSize
$text_secondary_fontWeight
$text_secondary_fontFamily
$text_secondary_lineHeight
$text_auxiliary_fontSize
$text_auxiliary_fontWeight
$text_auxiliary_fontFamily
$text_auxiliary_lineHeight
```

That's 33 tokens, less than three dozen to manage, which should describe a significant part of any experience cohesively in its resting state. You may identify plenty of design concepts that have yet to be covered by this set but to describe the ideal state of an interface, this should cover our minimal expectations.

To understand the benefit of our set of intents, let's compare a traditional token use against our newer understanding. Imagine in your system that you had a token simply called `$brand_primary_color`. You can imagine that this color is used for the logo, the primary button, and the background of some cards on your marketing pages. A new rebranding effort is required and you might think it is appropriate to simply change that color from teal to red to match the new style. However, you'll find that now the buttons look like they are all destructive, and the cards are very distracting. It becomes apparent that applying color this way is wildly inconsistent because "primary" describes the primary color to be used, which could be anywhere one feels like.

*Using the brand color haphazardly introduces
confusion about when and how to use it*

If we prepare our design tokens as intents instead, a design token name as a brand color will not exist because it is not a user experience concept. In the case of the logo, button, and card, they are all treated differently by using intents. We've already covered the button (action) and the card (surface, perhaps with its own contextual expression). So what is the logo? It is an illustration and we'll cover what that might mean in the next chapter.

Intent categories are meant to generalize the purpose of nearly every element that exists in graphical user interface design. These groupings of similar purposes are also a grouping of similar styles. We suggest the purpose of an element through its style. This is especially true on the Web since everything rendered in HTML is a rectangular box. It's the style of the box that suggests more about its purpose. Aligning the purpose to the style is precisely why an intent as a design token exists.

More properties and further tokens could be considered when making design decisions. So far, we've only considered color and typography styles as part of the intents. The next chapter will cover the properties of design decisions that are more appropriate for other intent sets.

Chapter Summary

We've learned about design tokens and their places on tiers in an expressive ecosystem. We've identified that intents are the most useful tier under a specific construction that describes their purpose, priority, and properties of element families. We've seen some examples of these tokens and how they relate to a few user-interface concepts that are common for digital experiences. In the next chapter, we'll discover the outliers that will improve our education on the system as a whole.

Key Concepts

- Design tokens bridge design decisions and their technical implementation
- Tokens separate visual decisions from structural components
- Three main tiers exist: primitive tokens, semantic tokens as intents, and component tokens
- Intents are the most important for maintaining scalable design systems

Token Architecture

- **Primitive Tokens**: Base-level values (e.g., `$color_teal_500`)

    - Should remain stable
    - Directly represent specific values
    - Used to communicate values easily between people

- **Intents**: Purpose-driven tokens constructed from

    1. Purpose (e.g., action, control, surface, text)
    2. Priority (primary, secondary, auxiliary)
    3. Property (background color, font size, etc.)

- **Component Tokens**: Should be avoided due to

    - Maintenance overhead
    - Risk of inconsistency
    - System bloat

Core Categories

Category	Purpose	Example Token
Control	Form inputs	`$control_backgroundColor`
Action	Interactive elements	`$action_primary_backgroundColor`
Surface	Content containers	`$surface_secondary_backgroundColor`
Text	Typography	`$text_primary_fontSize`

Best Practices

- Start with ~33 intents
- Maintain consistent priority levels
- Focus on user experience needs over aesthetic preferences
- Keep primitive token values constant
- Use intents to support different expressions of the same system

Common Pitfalls

- Creating too many component-specific tokens
- Naming tokens after specific values rather than purpose
- Introducing too many priority levels
- Using brand-specific tokens instead of intent-based ones

Endnotes

1. *https://lawsofux.com/hicks-law/*
2. *https://html.spec.whatwg.org/multipage/forms.html*

Outliers

In creating a design system, we aim for rules to govern the way
we maintain experiences. In the previous chapter, we covered
four categories of purpose which construct intents: "surface",
"action", "control", and "text". Unfortunately, there are outliers of
this approach that need to be addressed since they don't fit
nicely into the framework as described thus far. One of those
cases has to do with one more intent category to cover, and it is
the most challenging to communicate and further support in a
large organization.

Figures

The **figure** category covers amorphous regions that are nearly
identical in purpose but are designed to be visually distinct from
each other. Examples Include data visualization areas, stylized
illustration fills, and avatars that have yet to include user photos.
In these examples, the color indicates uniqueness against other
regions within the same context. In this way, each region shares
the same general purpose as its siblings, making our earlier
division exercise impossible because all of these regions exist
with the same purpose.

*Examples of regions that are
primarily differentiated using color*

I've called this category "figure" because it means to discern various regions using color or texture. However, it is accurate to assume that the colors you might want to use for data visualizations could be inappropriate for illustrations. This is where I'll admit a component token strategy is most appropriate. If you don't have any elements in your experience that expect this region coloring, then I am confident you won't need component tokens. If you have an experience where regions are existentially similar but must be presented as unique, consider the possibility of component tokens specific to this usage.

So if we are now considering component tokens for this category, what would the construction of these tokens look like? The best way to describe this exercise is like a color-by-number book. Imagine each region has a specific number to identify it; each number connected to some specific color value. One color for each number is the same way we assign a value to any design token. For instance, you could now consider a generic token such as `$figure_1_color`. The problem here is with curation. Before with the earlier semantic tokens, one could be fairly sure what would be affected by the given value; `$action_primary_backgroundColor` expects an assignment to the primary button's background. For `$figure_1_color`, the "1" must mean something. My recommendation is to represent a ranking for the numbers. Like priority, we could enhance the number to `$figure_1st_color`, representing the most common color we expect to appear for this figure. As a more concrete example,

`$dataviz_1st_color` would represent the first region to appear in any chart. The `$illus_1st_color` could be the brand color used to fill the largest regions of illustrations.

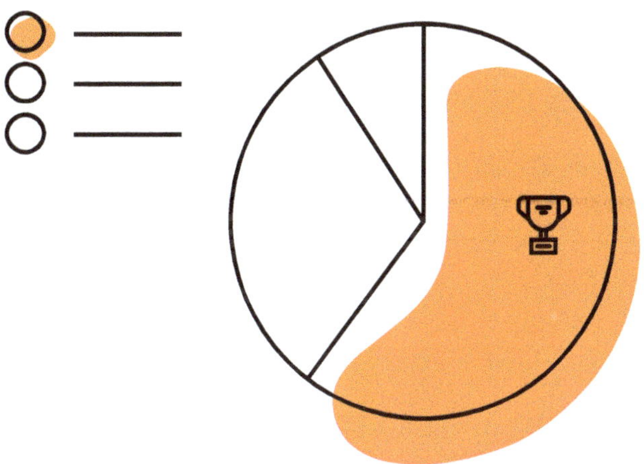

A data visualization using a
ranked color-by-number approach

The intents I recommend here introduce the component tier of tokens as an acceptable solution for figure elements. This is the case because the idea of "figure" is too broad for the concepts it is meant to cover. In other words, the purpose of data visualizations is fundamentally different from the purpose of an illustration but both of these concepts have a similar need to isolate regions by color. That's why they share the same general construction except for the identifying family of elements they are meant to affect.

The meaning of color value is much less important for other user interface elements, like avatars or even code syntax highlighting. To be clear, I'm not saying the color values chosen for these areas are not important. Instead, I'm saying what these colors are communicating is less important. As an example, the colors used for syntax highlighting should be chosen in an accessible way; no doubt about it. However, which color is chosen to convey variable names in the highlighting is less important. The use of color for these components merely suggests divisions in an otherwise homogeneous group of items. More examples of these could include user-generated tags for categorizing otherwise similar items and cursors for multiplayer experiences making it easier to identify each individual user.

The number of colors depends on how much needs to be covered. You could consider looping around the set in practice, as it could be necessary for dense data visualizations with areas of interest in the dozens. This is easily handled with the `:nth-child()` selector in CSS.

```css
path:nth-child(3n + 1) {
    fill: var(--dataviz_1st_color);
}

path:nth-child(3n + 2) {
    fill: var(--dataviz_2nd_color);
}

path:nth-child(3n + 3) {
    fill: var(--dataviz_3rd_color);
}
```

If only three colors were available, the 4th `<path/>` element would receive the first color in the set. The 5th `<path/>` would receive the second color and so on wrapping through these three colors.

While you might be inclined to include consistency in naming tokens by replacing the "primary" with "1st" across the semantic sets, it is important to remember that there is a difference in meaning here. In our earlier examples explaining intents, "primary" means the most important. In the case of the figure set, all tokens are of equal importance. In other words, just because something is first doesn't immediately convey that it gets the highest importance. The first color in a chart doesn't necessarily need to be the biggest slice, especially if the data is presented chronologically. So when we are curating these colors, we imagine where they might be applied, or better yet, look for an example of their use while curating and notating where color needs to exist to fully cover the regions appropriately.

As mentioned, this category is exceptionally hard to cover and curate because we are adding tokens based on a more specific component. A silver lining here is that you don't need to introduce these unless you have something that needs to fall into this category. If your experience doesn't need partitioned regions of color for anything, then feel free to ignore this category for the time being.

Interactions

You may have noticed that so far I've avoided "interaction" concepts like "hover" or "dragging" from the construction of these tokens. While these could be easily included into the sets, I recommend avoiding these when possible for alternative approaches. The main reason is to keep the number of tokens you maintain low. However, another reason has to do with the importance of curating these interactions with new treatments.

Let's think about "hover" as an example. For most digital experiences, 50% of visitors are using a touch device which cannot present the concept of "hover" to users. Furthermore, the treatment of an interactive element should convey the possibility of interaction before a person even considers moving their pointer to the element. Once they confirm that the element is interactive with "hover," the user feels better that they've guessed right. We can provide other treatments such as changing the cursor from the default arrow to the hand pointer on hover. Using the cursor as an indicator is better than a color change for those with poor vision. This doesn't mean you can't affect color. You could use the current background color provided by an existing intent and use it to change the background slightly when the element is hovered. This avoids the need for curating a new color, which avoids the need for a token to describe it, in turn keeping the number of tokens as low as possible.

```css
button.primary:hover {
  /* Using relative color syntax */
  /* Lighten teal keyword for hover background */
  background: oklch(from teal calc(1 + 20) c h);
}
```

In the case of "dragging", you can consider this as a change of surface from an auxiliary priority to a secondary priority once the user grabs the element. This makes sense because the user is currently focused on this specific surface separate from other levels of surface priority that are composed in the interface. The surface currently being dragged is temporarily more important than the sibling surfaces that are not being dragged.

In the case of "focus," I recommend you avoid curating this entirely. Users who rely on this feature have expectations of its presentation, and taking control of this will now make your organization responsible for maintaining those user's expectations along with whatever accessible customization they need which could easily be in conflict. Consider this as a critical way of navigation for many folks. You don't want to mess with it. It's a similar level of effort to creating a new accessible pointer cursor, so I recommend leaving that as the default.

Now if these options aren't available due to pressure from the organization, I recommend the following construction:

`$purpose_priority_interaction_property`. There's a few reasons why I suggest this construction over others.

First, this will help with tooling later down the road. It is easier to group all of the background colors together if you can simply look at the last entry in the construction to determine its type. In the future, if you need to create an interface to help supply values to these tokens, you can tell the kind of value expected by reading this final part of the construction. This could help inform the component needed to input the value. For example, seeing "color" at the end would require the use of a color picker for the value.

Additionally, it follows a similar pattern to how this would have been written for CSS:

```css
/* Selectors become more specific from left to
right */
button.primary:hover {
  /* Property is last */
  background: ...;
}
```

> ### ⓘ NOTE
>
> I'm writing `background` instead of `background-color` so that the token may support gradients as a value.

The token used as the value in the example would be `$action_primary_hover_backgroundColor`. You can decide if you should use "hover", "hovered", "hovering", or some other tense of interactivity here. You'll want to decide on this upfront going over all the possible interactions you'll want to cover. This is supposed to be a system you're making after all.

However, in keeping with the goal of keeping the token count low, I recommend avoiding this exercise altogether if you can since they are often fleeting presentations.

Expressions

The final topic in the world of outliers we'll need to cover is **expressions**. An expression takes an existing representation of the experience and alters its presentation to convey a new idea or message. You can think of this as introducing a new wallpaper to a room to add some personality. Maybe it's totally 80s, mid-century vintage, or pure minimalist, but the construction and purpose of the room remain the same. Making this change suggests a new feeling for the person experiencing it.

For example, let's imagine a form field expecting an email address. If the user enters an invalid email address, we need to change the input to express that there's something wrong. If you've dealt with design tokens in the past, you most likely created tokens to describe the concept of feedback to support this example. When the form is in an error state, we replace the default tokens with ones that have values conveying the concept of being wrong.

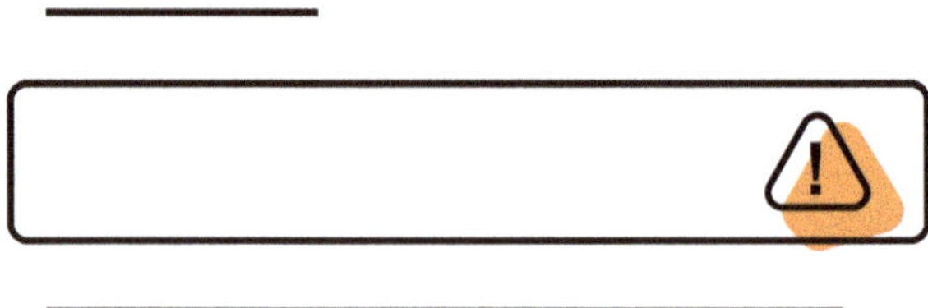

An email form field with an error should
express a different looking state

Something should sound weird to you here. We're replacing one variable with another variable. Variables are meant to hold values, and importantly, those values can change without needing to affect the variable's use. So, instead of replacing a variable name with a new variable and associated value, we should *redeclare* the existing variable with a new value. Redeclare means there was one value assigned to a token, and we're going to replace that value with a new one. This doesn't change the name of the variable, only the value that was assigned to it. In this way, we never need to revisit where the tokens are used. Values are delivered differently depending on a newly introduced set of assignments.

Let's see what this might look like in CSS. Here's what you may have done before:

```css
:root {
  --control_borderColor: gray;
  --control_critical_borderColor: red;
}

input {
  border: 1px solid var(--control_borderColor);
}

input.is-critical {
  border: 1px solid var(--
control_critical_borderColor);
}
```

Here's what I'm suggesting:

```css
:root {
  --control_borderColor: gray;
}

.is-critical {
  --control_borderColor: red;
}

input {
  border: 1px solid var(--control_borderColor);
}
```

While these approaches are different, both of these would render identical presentations. You should notice a few things in comparison. First, it's less to write as code. But it's also less to maintain as tokens. There's still two values that need to be set, but in the former example, there's also two tokens to maintain. In the latter example, there's a single token to maintain but it gives the same result as the first, so long as the input is within an element with the `is-critical` class name.

This means that we aren't restricted to conveying the idea of being in a critical state to inputs. We can convey critical information for other categories such as surfaces or actions. Imagine that your alert doesn't have new tokens, it has new values to existing tokens. This is why I believe there is no such thing as a destructive button, only a button expressing the idea of destruction. We can convey the idea of something being critical just by changing its surroundings and never touching the elements inside.

The way that I identify when to consider a treatment as an expression is to start by identifying if the presentation can be defined using existing intents. This exercise isn't trying to consider what the values for those intents are. We are reviewing if this case could be described with our existing tokens. As in, "is this considered a surface?" Good, we have intents that describe a surface. "Does this have text?" Good, we have tokens for that too, even though the final presentation could be cosmetically different from a change in state. If we go through this exercise and find that no new concepts are being introduced, this suggests that the new treatment can be considered as an expression instead of adding new tokens.

A great example of this is the concept of "selected". Again, it is common for folks to assign new tokens for this concept, but it breaks down in more complex compositions. Consider an interactive table where each row has actions a user could take on the data. If you have the ability to select multiple rows, each row will need to convey the concept of being selected. If you introduced new tokens for the selected background color, that might not work with the controls that exist on that background. You'd need to introduce new tokens to describe those controls on this differently expressed background. However, if you introduce a new expression to the entire area, you can assign new values to the existing tokens for all of these elements, such that they all work together within the selected expression. Remember, if it's less than three dozen tokens, it is very manageable to decide values for all these and get full coverage no matter what element appears in a table row.

How could we possibly prepare tokens for all the kinds of elements that could appear in a selected table row?

You should now begin to see the power of this idea, to use the existing tokens and change the values based on the expression we want to convey. It keeps the maintenance low while opening up the experience to lots of different expressions.

> ⓘ **NOTE**
>
> You might be thinking about the complexity of combining expressions. For example, the idea of "selected" in a light-colored page being visually distinct from a dark-colored one and requiring a new presentation specifically in this area. We'll address this in later chapters. For the moment, focus on each expression in isolation to keep things simple. We'll ramp this idea up over time.

While we've focused on color for demonstrating the concepts, there are other stylistic properties we'd like to affect. In the next section, we'll explore more about translating intention into design tokens for enhancing our experiences.

Drawbacks of Overgeneralization

A conclusion that some folks come to when considering this architecture is "why not have *everything* be a surface? Isn't a button just a more specific surface? Couldn't we reduce the number of tokens more to only include `$surface`?" The answer is yes, you could, but at the cost of requiring more expressions.

As an example, you should be very confident that your experience will include a primary button. In this overgeneralized architecture, you'd need a new expression to describe the concept of a primary button and apply it to every element meant to convey this to the user. That button wouldn't have any other defining qualities to identify it as a button until the expression was applied. It is simply a surface waiting to be something more.

To make things more complex, that button could exist within a light or a dark mode. This changes the surroundings where the button is placed. This means you'd need to have expressions that were more existentially aware of these environments. In other words, you'd have to have some expression that conveys surfaces as "light button" and "dark button" which would have to be used in conjunction with the more generic expression concepts of "light mode" and "dark mode" user preference. It is simply easier to include the concept of an action or control within our set with a purpose more significant than being a simple surface.

The takeaway here is that we don't want to give an element meaning by supplying an expression, we want the element to have a purpose before we give it an expression. The goal is to create a new expression for the element that was placed there, unrelated to the reason why it is there.

Another way to think of this is about placing elements on a surface. We should be comfortable with the idea of text and icons being placed upon a surface, how the colors of those glyphs should be associated with the surface they were placed upon. We should extend this to having actions and controls also placed upon a surface. This will keep a relationship between these elements and the surface they are placed upon. In fact, in the curation exercise, keeping the colors of the current `$surface` elements in mind while providing other colors to intents is a good practice. For folks familiar with the way light and dark mode are traditionally curated, this should match an existing mental model.

Importantly, being slightly more specific in our surfaces doesn't stop us from placing a button within its own new expression, if a designer chooses. We have already explored this possibility by replacing the "critical" button with a button within a critical expression. How large the scope is for "critical" is up to the designer. The expression could wrap individual elements if it's appropriate. However, we'll want to do our best to increase the scope to show a cohesive expression among elements, not provide unique treatments on an individual element basis.

An alert is expressing something different from the rest of the page, but the basic elements inside remain the same

Chapter Summary

The practice of design systems is no stranger to snowflakes, but identifying them early can help get ahead of the ambiguity they tend to promote. Having a clear structure to the system and reasons behind the decisions makes outliers feel more a part of the plan. Opening up possibilities of expressing ourselves in ways we haven't done before.

In the next chapter, we'll explore how to identify which visual properties can be effectively managed through our token architecture, and more importantly, how to recognize when certain properties require different approaches. This exploration will help you make informed decisions about extending your design system beyond basic style properties.

Key Concepts

- Figure elements require careful consideration of component tokens
- Interaction states can often leverage existing tokens
- Expressions provide powerful customization without token proliferation
- Overgeneralization can lead to maintenance challenges

When to Use Each Approach

- Use component tokens for figures when regions need distinct identification
- Leverage existing tokens and CSS features for interactions
- Apply expressions for state-based interface changes
- Maintain specific intent categories to avoid overgeneralization

Best Practices

- Keep figure token sets minimal and purpose-specific
- Avoid custom focus states
- Use expressions for state changes instead of new tokens
- Consider scope and composition when applying expressions

Identifying Intentions

When considering design tokens, color and typography are largely discussed because making one change here can have a large visual impact, depending on what was changed. For example, updating the `$surface_auxiliary_backgroundColor` can really cause a stir. Other kinds of visual changes are often less noticeable or harder to create a system around.

Of course, there are plenty of other possible styles that can influence an experience. We could introduce configurations that affect the orientation of child elements, or the visibility for example. However, the key concept here is that those kinds of decisions are very likely to change the purpose of what is presented to the user. This kind of update could fundamentally alter the functionality of the interface, good or bad.

Encoding intentional changes into tokens is no easy feat. There's a lot to consider. Perhaps more than any one designer is meant to handle themselves. This is one of the main reasons why we typically simplify visual changes to cosmetic ones such as color and typography.

This isn't to say other design decisions can't be represented with a truly intentional design token. Let's take a look at some other ways intents could support customized needs and where they can't work by definition.

Density

After color and typography, the next most popular category for user interface design style is **density**. Density determines how tightly packed the content appears and is directly related to the amount of space between elements, among other stylistic qualities. It is common for data-intensive interfaces to be very dense with information because the user is expected to process a great deal of information simultaneously to continue further steps. Meanwhile, marketing pages are much less dense, so it is obvious what the user should do next with lots of space in between sections. Less information is present on the page, and there are fewer choices to make. Keep these examples in mind as we further consider density.

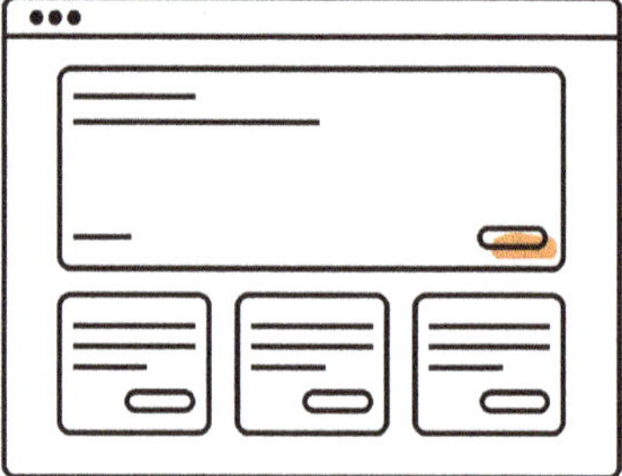

Comparison between a dense dashboard
and a spacious marketing page

In our current mindset, we can see that the amount of space between the elements does not necessarily change the layout unless pushed to extremes. This makes density a candidate for being an expressive design decision. As suggested earlier, more space could help lessen the choices. Less space could help a user make a complex choice with more data.

In many public design systems, space tokens are provided with a scale. We've seen scales with color tokens in earlier examples, such as `$color_teal_500`, and have already identified a scale like this as part of the primitive tier. In fact, some examples of space tokens are even more descriptive than color, explicitly writing a proportional value within the token name itself, such as `$space_4`, which is commonly meant to represent a space that is four times as large as a base `$space_1` token. As we've identified earlier, intents do not depend on each other and allow regular systematic insertions. Needing a space token between `$space_1` and `$space_2` would make for a cumbersome naming convention. The goal here will be to curate amounts of space while avoiding the restrictive qualities of primitive scales.

To do this, we must understand why we add space. Not for a specific experience but from a universal design standpoint.

The Gestalt Principle of Proximity[1] states that we perceive objects closer together as related or as a group. This means when choosing an appropriate amount of space, we are really grouping elements to indicate a relation. When there is no space between items, they are meant to be very closely related. Conversely, they are the least related when a great deal of space exists between them.

Another important note about space, famously described by Albert Einstein, is that *space is relative*. For example, the distance between you and your kitchen is significantly shorter than between you and the moon. However, the distance between you and the moon is significantly shorter than the distance between you and the sun. How could the distance between you and the moon be large and small simultaneously? It depends on the observer's perspective. The size and space between objects aren't changing; only the scope of observation changes.

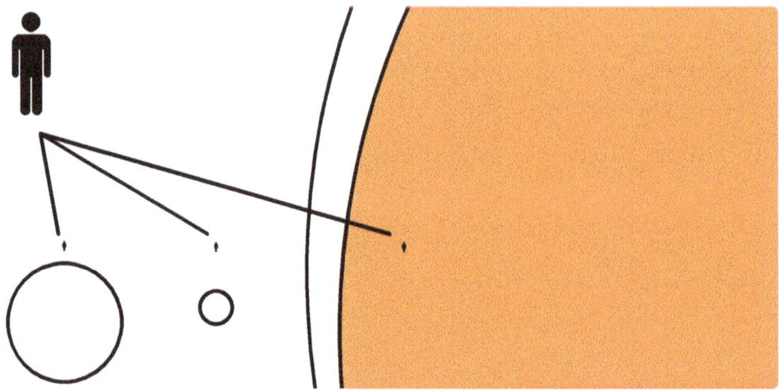

*Comparison between a person
on Earth, the Moon, and the Sun*

Another way to think about this is as a dollhouse that is a perfect replica of your home. Imagine we used a distant telescope to look at your life-sized house. The scope reduces our reference to only what we see within the confines of the device. Now, imagine we move to an appropriate distance to view the dollhouse within the telescope so that the image looks identical. Because we have no frame of reference, we cannot tell the size of the objects. Our perspective from outside the scope shows differences in size and scale, but it's all the same inside the scope.

A house interior in the viewfinder of a telescope

With this in mind, we can reconsider spacing tokens as an intent by reducing their number to one single token. This sounds extreme, and it is. Bear with me.

Imagine placing a single spacing token in your interface at every location where a relationship between items should be provided. Again, these aren't differently named tokens with a scale; this is the same token perhaps called $space. Imagine we used that telescope to zoom into different interface regions; each time we zoom, it represents the amount of space to be applied to those tokens. In this way, we uniformly set the value to all tokens within the visible scope and do not affect anything outside that scope. The number of zooms equates to the multiplier of space or the number of density levels appropriate for this composition.

In practice, the scope I'm introducing is not an actual telescope but a defined region in the composition that is a boundary for new values. Intents across the composition may be given a different value based on the scope within which they are found. For example, imagine an informational table. In traditional user-interface design, we may provide a variation of the table component that is specially updated to support data-dense needs. In this new paradigm, what if this table component's density was influenced by the scope within which it was found? This would mean that a person developing an interface would not need to configure the option for it to be dense. It would be dense purely by its very existence within the denser scope.

As a complete system, the largest and least dense scope would exist at the page level, and each time we need to increase the density, making things more compact, we introduce a new scope. This makes sense, as smaller items within a larger composition have less space for content within as we nest elements. For example, comparing the available space for a big chunky hero section to the space found within a single tight card element. The depth controls the scale of space and we introduce levels of depth with nested scopes. This removes the decision for which token to use or how it shall be applied. As a bonus, we can also avoid a naming exercise entirely if the space is simply reduced each time a new scope is introduced. Introducing a new scope is enough to require new values, which could be provided at a regular interval. The difference here is we aren't introducing new ways of delivering the value. This would be like zooming on a telescope by clicking a dial over picking up a new telescope locked in only one zoom level.

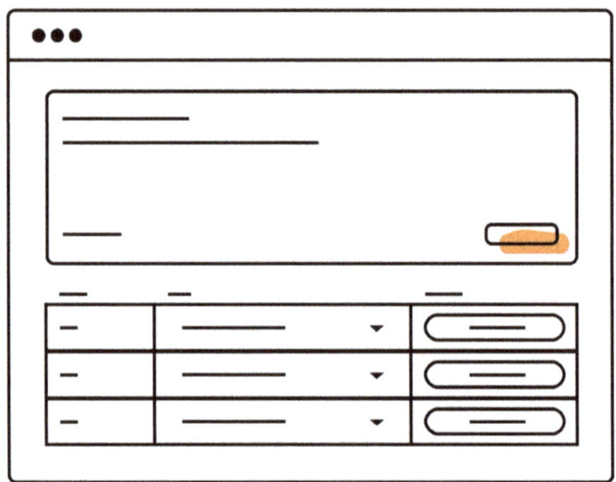

Nested levels of density changes in a layout,
from the marketing hero to a dense pricing table

Your next discomfort may be that there's a lot less control than one would have when setting primitive values explicitly. That's true, and that's also the point. If we think about how we are meant to handle color in an intent-forward system, we might apply `$surface_auxiliary_backgroundColor` to the page. This technique doesn't permit the designer to control that value precisely when placing a token in the interface. Placing that token is about understanding the purpose of an element and then allowing style values to be applied from some outside influence. The decision about what value is assigned to `$surface_auxiliary_backgroundColor` occurs in a different set of responsibilities, as this could be a light color or dark color depending on user preference, for example.

The same is true for space in this way, where a user experience designer's decisions will evolve. They will now decide where space should go between elements and where a containing scope change would occur. Combining these decisions, plus values set in a separate token curation responsibility exercise will inform the resulting space. In our telescope zooming dial metaphor, this is the separation of responsibilities where the manufacturer creates the dial and determines the level of zoom each represents. At the same time, the user can set the zoom to any one of those predetermined notches.

Another important point about considering this approach is that the space token we're identifying is the space between objects. This space token may be less appropriate for contributing to an object's dimensions. For example, the padding applied as a signifier around a button for the interactive target area affects the element's size, which could differ from the amount of space between adjacent buttons meant to show a relationship. To clarify for a larger organization, you could provide the space tokens for `$padding` and `$gap`, which would suggest they are meant for space around and between elements, as we've described.

We've learned here that the value of an intent can be updated based on where the intent is used. This opens up the possibility of creating a highly curated set of values for a specific spot in the experience without introducing any new naming or resources. This unlocks the possibility of infinite values with a small number of tokens. Very powerful stuff!

Roundness

Another stylistic property that often adds personality to an experience is the corner roundness of containers. Because this quality does not affect the overall purpose of an experience, we can assume this may be a candidate for an intent over a primitive scale to provide a new expression.

For this exploration, we'll assume that the rounding should all be uniform, and no selective rounding is applied on any individual corner. For example, this would mean that rounding treatments often used for chat bubbles to indicate message ownership are out of the scope.

Let us consider our token naming schema. Is roundness expected to convey priority and include qualifiers such as "primary" within the name? The answer is no; roundness is not a signifier of priority. Elements do not become round to draw attention to them compared to similar elements. At best, we could stretch to say that the sharpness of unrounded corners conveys a harsher meaning from a rounder one, but these are still simply expressions so far.

Our next question is: should rounding values be category-specific? Do we need to allow rounding to be different between interactive elements and their containing sections? What happens when we are deciding how round something is supposed to be?

A seasoned designer would suggest that the amount of roundness for a card and a button within the card should be different but directly related. In other words, we often aim to achieve concentric corners; *perfect nested rounding*. In technical terms, this occurs when the circles that would contribute to the rounding of both elements share the same center point. In execution, to know the amount to round one corner, we need to know the roundness of the closest corner and the distance between these elements' boundaries. That distance is the amount needed to offset for that perfect nested rounding.

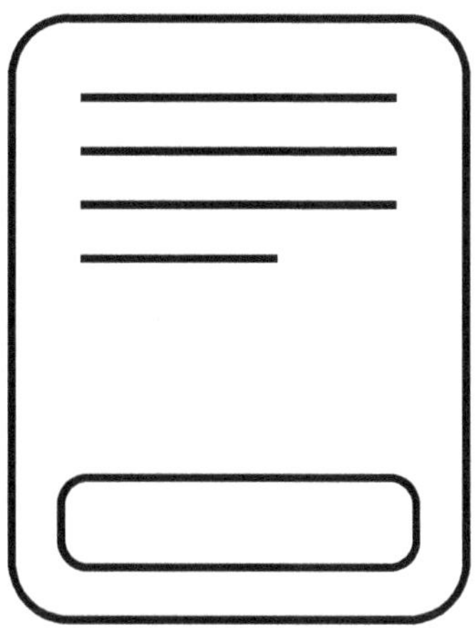

*The distance between the button and its container
determines the perfect corner roundness*

If the way in which rounding should occur is known, then shouldn't whether something is rounded or not effectively be a binary decision similar to our $space example above? Can we move on with something like $curve where it is simply applied where we expect this behavior? While this seems straightforward with a known calculation for the desired effect, using it as a repeatable treatment is deceptively challenging.

For example, if this depends on the roundness of the closest corner, what does that corner's roundness depend on? Its own closest corner, of course! This highlights that there must be a starting roundness for all other roundness to be derived. Where does the derived rounding start, from the largest or smallest amount? If we start at the largest and work our way down, we could theoretically reduce the rounding so much that our deepest corners would be zero, with no rounding at all. Starting at the smallest and going up, we may curve so much that parts of the interface become comically circular.

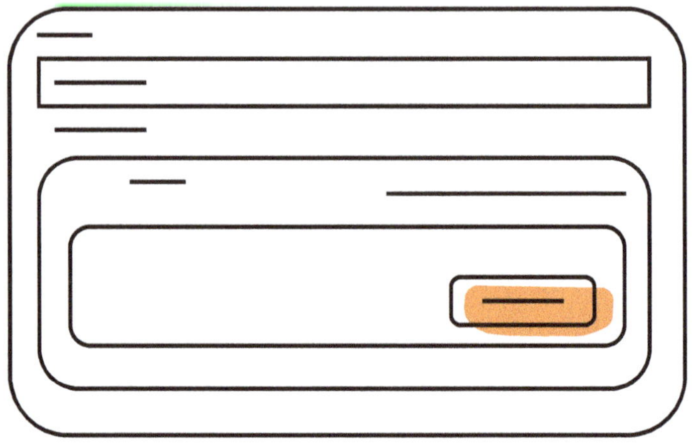

*If rounding happens from inside, elements without
rounded children will also not receive rounding*

OK, maybe we can apply minimum and maximum thresholds for
rounding. Let's consider a group of three adjacent cards. One of
the cards includes a button, and the other two do not. Certainly,
we'd like all the cards to appear uniform as siblings. However, if
we use the bottom-up method, the button would influence the
rounding of only one of the cards, and the other cards would not
receive the same influence.

*If only one element has something
rounded inside, they become inconsistent*

Meanwhile, in the top-down method, it is possible that a navigational bar spanning the width of the interface is meant to receive no rounding, so smaller elements like buttons within would have no influence from the parent and, therefore, also not receive rounding.

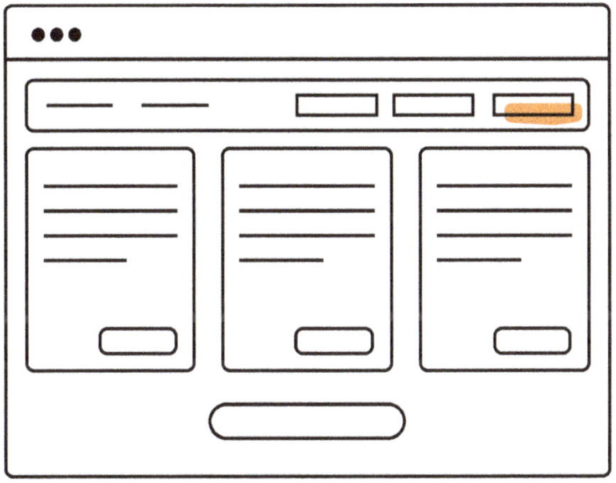

Rounding from the outside-in may result in elements that have no rounding at all

Thinking deeply about this subject, it becomes clear that perfect nested rounding isn't possible to describe as a binary decision due to all of the layout factors that are affecting the desired result. Because those layout factors are primarily functional in their placement, we cannot apply this value in the same way as `$space` above. The best we can do is apply rounding based on an existing intent construction. This would introduce tokens like `$action_primary_cornerRadius` meant purely as an expression of roundness and not as a way to achieve the perfect nested roundness.

What we've learned here is that the treatments we can affect with an intent are isolated. They cannot depend on other factors outside of the value and scope where values are supplied. In the case of perfect nested rounding, the value and scope aren't enough to present an appropriate result so we can't support the concept purely through this approach alone. Meanwhile, if a marketing department expects the overall experience to be "more friendly", supplying rounding as a general treatment is achievable.

I have confidence there is a way to achieve this from a systematic perspective. But since this is tied to layout, an answer is outside the scope of this writing. Research for another day and another book.

Express Yourself

We've learned some important qualities of intents which can be used to inform other properties in an interface. This means we can consider this exercise for other possibilities. In this section, I'd like to take the opportunity to go through my thinking for intents to describe motion so that you could go through similar thought exercises when considering how you might allow other cosmetic properties to be expressed.

⚠ **WARNING**

Before going on, it's important to disclose that I've never completed a thorough analysis on how I'd introduce motion into a design system before this writing. What you're about to read in this section is truly my initial thoughts into how I'd consider adding motion without any previous research or development. In my projects, I typically avoid motion due to its distractive nature both in an experience when overdone and from the curator's point of view of choosing values. The need for making many decisions largely based on feelings often without understanding intent. In a fully realized addition to an existing set of intents, it would take a great deal more consideration into how motion fits into the larger ecosystem. This is only a taste.

The first question I'd ask is: *why* add motion to our experiences at all? The largest reason is due to **change blindness**. This is when something in our purview changes without being noticed. We can use motion to help draw attention to a change over time. We could consider that smaller changes need to last longer than larger, more noticeable changes. I don't necessarily mean that the element itself is a large or small size. Instead, we should consider the size of impact in a similar way to how we think of priority.

This requires us to consider what is expected to change in an experience. Should a button ever change within the lifecycle of an experience? The answer is complex, because the button should have the same purpose and priority throughout its existence. However, it is very common to show a slight change when performing a minor interaction with that button, such as hovering with the cursor. That could introduce a transition from one color to another. The impact of such a change is small.

The next thing I'd consider is the difference between functional motion and expressive motion. This aligns nicely with our previous separation of concerns. Let's consider an animation of a clock ticking down. This will illustrate a good test to identify the difference between motion that is functional versus one that is expressive. A ticking clock moving at a regular interval is functional movement and conveys the normal passage of time. Meanwhile, a clock where the hands move irregularly will convey a different message since it no longer maintains an expectation of time. Instead, maybe it is meant to describe something more fun and playful. Even a clock that doesn't move at all can convey a message.

*A clock ticking away conveys a different
message than one standing still*

So let's consider something more common in today's interfaces. What kind of motion could be added to a panel opening from one side? Certainly, a panel appearing suddenly could be startling, so the motion is helpful to gradually introduce the content. Again our question is, if we were to make this motion configurable, would that configuration be meant to change the message that the panel is meant to convey? In my opinion, it would not. The panel is largely a vessel for content; how it enters and exits is more expressive than functional. A person can certainly have feelings about the rate at which it transitions, but ultimately the way in which it enters and exits does not change the functionality of the panel—the reason why it was placed in the experience.

Something else to consider, unique to motion, is that motion could affect any presentational property. We've already seen examples of motion affecting color and panel sliding, but we could also consider other properties like opacity or growth. This is where we should prioritize what would be the most helpful to control from an impact perspective. The way in which a surface enters and exits an interface could support a more expressive behavior. Less important surfaces might call for a more expressive transition to be noticed.

One of the final considerations is what properties you expect to expose as values to be curated. We want to maintain a small amount to avoid overwhelming a person choosing the values for intents, but we also know there are several factors that go into composing motion.

My recommendation here would be to allow the duration and easing function but not allow a configuration that directs how those values are applied to a pattern or component. This means when someone supplies the timeline of motion, what it affects is at the discretion of the owner of the user interface element. This isn't any different from other intents. A person building a component could choose `transparent` as the only "color" meant to be applied at a location, effectively meaning no color. They can also choose that the `foregroundColor` is meant for the `fill` of an icon. The same goes for choosing not to include an animation or choosing how those values contribute to the experience.

This means I would consider introducing the following intents:

```
$surface_primary_motionDuration
$surface_primary_motionBehavior
$surface_secondary_motionDuration
$surface_secondary_motionBehavior
$surface_auxiliary_motionDuration
$surface_auxiliary_motionBehavior
```

Where `motionDuration` is the length of time the motion should play and the `motionBehavior` is some timing function that produces a Bézier curve controlling how the motion is meant to behave over that duration. Again, *how* it gets applied within the experience is up to the expectations and guidelines of the system, most likely from inside reusable components.

Chapter Summary

In this chapter, we've explored other stylistic properties outside of color and typography that could also contribute to the overall expression of an experience. Density can be changed at different scopes, while roundness is more closely tied to the structure and composition of the layout. We've also explored considerations for intents in other more nuanced properties like motion.

In the next chapter, we'll introduce the concept of modes—a powerful approach that allows us to change how our tokens are expressed based on context. This will fundamentally change how we think about managing design variations, moving from a token-based approach to a context-based approach.

Key Concepts

- Not all design properties can be handled the same way in an intent-based system
- Some properties (like density) work well with scoped values
- Other properties (like roundness) are deceptively elusive
- Consider these questions when adding new properties:

    1. Does it affect functionality?
    2. Is it scope-dependent?
    3. Can it be expressed without component-specific tokens?
    4. How will it scale across the system?

Endnotes

1. *https://lawsofux.com/law-of-proximity/*

Modes

So far, we've introduced the possibility of changing the value of an intent based on its placement within the experience. This was demonstrated when conveying a new expression by enhancing feedback messaging, and also in the case of presenting a more data-dense interface. In this chapter, we're going to dive deeper into this concept.

A Whole New Scope

In the section describing outliers, we reviewed a new way of thinking about how we want to express the idea of an input being in a critical state. As a refresher, traditionally we'd create new tokens to describe the critical state as different from the resting state. When that critical state occurs, we switch out the tokens to present the new value. Our new mental model says we don't switch the tokens themselves, but switch the value that the token represents.

We can achieve this by considering that this input is different from all other inputs at this moment. Traditionally, we'd possibly need to update the label and the text underneath in a new cosmetic treatment to convey this. However, we could also allow these parts to be influenced by their location. This does not mean we move the components to a new place. Instead, we update what it means to exist in their current scope.

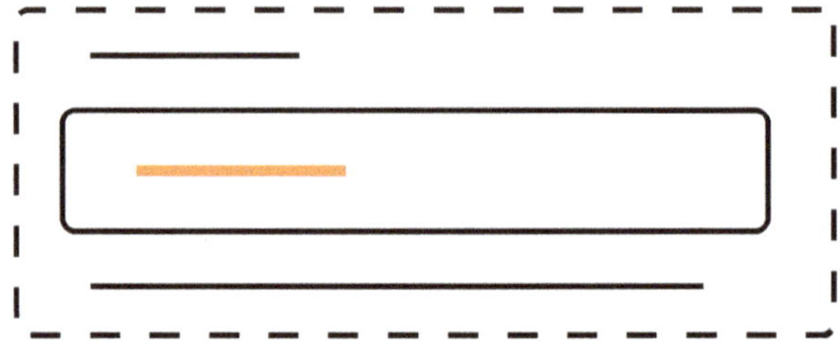

The scope of the input and items related to it

By updating what it means to exist within this boundary, we can present a new idea. In our current example, we'd only change the meaning of the current scope from default to critical; all the elements within would receive new values based on the newly introduced scope.

As another example of this technique, let's consider a banner meant to alert a user of something dangerous that might occur if they don't act soon. In a traditional way of preparing this, we'd have tokens that would describe a colored background and the text that appears on top of it. This is fine when the content is only text. But if the content needs to be interactive, we can run into color conflicts. What do inline links look like? How about secondary buttons? Maybe you allow a person to input the change right in the banner. What does the field look like on this colored background? There's a lot to consider when making new tokens to support this. This also doesn't yet account for the other possible kinds of feedback that an alert could support that aren't as important.

Lots of possible elements can appear inside of an alert

We could opt not to color the background so that it uses the more standard `$surface_primary_backgroundColor`, making this message look like any other meant to disrupt the user. However, one could say that the messaging needed here isn't like any other. It is meant to convey something more important. So how do we support this within our current constraints and mental model?

Let's consider this alert banner and how not just the button but all of the content within is meant to convey the concept of a critical expression. If this pattern is meant to introduce friction so that the user gives a second thought to the next action, it would be appropriate to also make changes to other elements within this scope to further convey the new importance. Perhaps colored icons and borders that are given a different treatment from those found elsewhere across the interface would show the user that this messaging is different from the rest. We could consider that this alert is presented in a "critical" mode, potentially influencing the presentation of elements within.

Previous chapters have suggested that there is a scope in which values can change. In the density examples, we changed the zoom level of a telescope to make more space appear between objects. In our daily design practice, we can introduce a new context that changes a token's value when we want to reduce the space between elements. That scope is a new **mode** meant to present a new expression.

You may have heard of light mode and dark mode before. These describe a user's preference for an experience in terms of color, where the values of tokens are curated so that everything is meant to be presented with a lighter or darker appearance. In light mode, elements are designed to exist primarily in an interface consisting of a light-colored background. In dark mode, elements are prepared to exist on a dark-colored background. Users might choose either preference for several reasons, ranging from readability to performance and even simple unconscious design attraction. Supporting these preferences is a good user experience design practice. The more comfortable a user is with an interface, the easier it will be for them to use it.

Imagine adding other modes, like one that sets the elements closer to each other making more information visible within the viewport; a "compact" mode. Maybe one that expresses promotional content; a "promo" mode. As you can see, this isn't just a matter of user preference, it could be another design decision meant to enrich the existing experience. A mode can enhance an otherwise unexpressive experience by curating new values to represent a new look.

A compact mode can help present lots of data, and a promo mode can help focus on selling features

Importantly, notice that during our exploration, I haven't restricted this concept to describing only a page but also sections of a page, like a modal. This means a mode could be assigned inside a smaller page area. If we consider the page to have "Mode A" and allow for sections of the page to each have "Mode B" and "Mode C", we are placing multiple modes within a mode. It's mode-ception!

This is the reason for the title of this concept as a whole: **Mise en Mode**, which means "placement in mode." It derives from the art history term "Mise en Abyme," which was the act of placing smaller works of art inside of a larger composition. In Mise en Mode, we place smaller modes inside of a larger one to change what we intend to express as needed.

You may have reservations about what is being described as a "mode", where you may have found the word "theme" used in some practice. The primary reason for this will be more clear in later chapters, but technically there is a difference between a mode and a theme. A mode is a presentation from the user's perspective as a concept. Meanwhile, a theme is a presentation from the designer's perspective by curating values. Another way to think of this: a designer creates the theme to support an expression; a mode. The user selects a dark mode and expects to receive a dark-colored experience, curated as the dark theme. A presentation may be provided to a page without user influence which would be more accurately identified as a theme, such as a special treatment for Halloween.

For our explorations, we'll be using the word "mode" for both purposes interchangeably. As an example, if we are going through an exercise of curating a mode we are truly selecting values for a particular theme to be affected by a mode. The separation will be more useful when discussing how to execute this in practice. Ultimately, we shouldn't mind what values are being assigned to create the new expression. We should focus more on the overall approach, and that is supported by thinking in modes.

As a plus, the alliteration of the phrase "Mise en Mode" I find subjectively delightful.

Avoiding Inverse

A way to demonstrate how we might use a mode in place of other traditional approaches is the common design choice of a light-colored page with a dark-colored footer. Traditionally, we might achieve this with design tokens by introducing a naming convention that includes inverse so that the same elements appearing in the page's light part could also have variant tokens for when they appear in the dark part. Perhaps the team has introduced the default light token as `$section_default_backgroundColor` and its inverted partner `$section_inverse_backgroundColor`.

A light-colored page with a dark-colored footer, marked with inverse tokens

This seems reasonable for simple compositions with text or a button within the dark footer because we're only introducing a handful of new tokens. Now imagine adding an email input for a newsletter in this footer. You'd need to create an input variant to also support inverted which would require new tokens. Now consider when that input is in an error state; you'd need to include even more inverted token variations. This could become hard to maintain as with any new component that we weren't prepared for; we would need new tokens to describe inverted properties. This could eventually result in an entire duplicate system of tokens, specifically when these elements are within an inverted background. Every token found in the "default" would need an associated inverse version with a new value attached. This is simply not desirable from a maintenance perspective. The amount of tokens you'd need to curate and later maintain would become unwieldy for larger organizations with many different design treatments. In other words, a design system maintainer would need to figure out the naming convention for these tokens, and a design stakeholder would need to assign new values for those tokens. As we said earlier, the job of maintaining and the job of curating should be separate.

Using what we've learned so far, we can continue to support the possibility of a light page with a dark footer without exploding the number of tokens we need to use. Instead of thinking of the footer as inverted requiring new tokens, we can think of it as a dark mode being applied within the scope of the footer. In a well-crafted system, we'd expect that dark mode has values assigned for all tokens meant to cover the entire experience. Some dark color is needed for `$section_default_backgroundColor`, and some light color is needed for `$section_default_foregroundColor`, with all other tokens also covered with values. This means there shouldn't be any surprises when introducing a new component, as it would have already included a dark mode consideration as if it was meant to be applied to an entire page. This was the exercise of naming intents, describing the purpose of an element and not its presentation. That exercise aims to cover all possibilities that appear within an interface, as completely as possible, in a conceptual way.

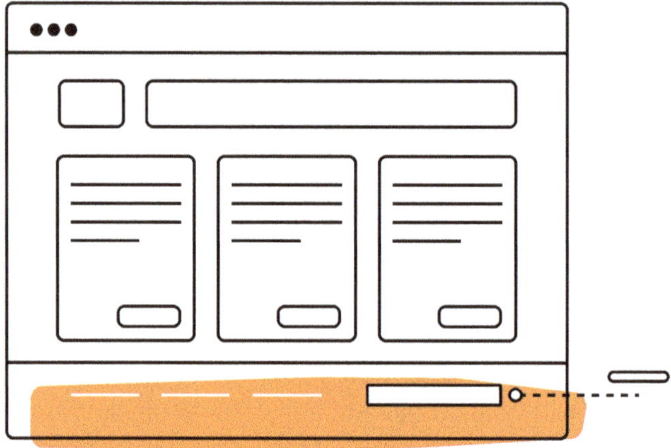

A light-colored page with a dark-colored footer, marked with modes

You may find that the values curated for the page-level dark mode are inappropriate to use as the footer-level dark mode. Perhaps the footer is meant to be a more stylized dark mode which uses gradient backgrounds instead of solid colors. In this case, it would mean that you are maintaining at least three modes: a light mode meant for the page, a dark mode meant for the page, and a dark mode meant for the footer. This is the weight of choice to enhance expressively. The footer doesn't need to be dark; it should function properly without that treatment. A designer may provide an enhanced presentation of the footer and choose a separate collection of values for this purpose.

At this point you might begin to feel overwhelmed by the number of modes an organization might need to manage. As in the example above, we'd need three modes, and one of those is specifically made for a single component's presentation which might not ever use the new values. It's true that moving the responsibility of new values from tokens to modes isn't lifting the weight of curation. It is still the same number of values that need to be curated. Sometimes even more than before, when you include expressions we expect such as "critical" and "selected" as a mode instead of new tokens. It's more than possible that the experience never sees the critical version of `$surface_secondary_foregroundColor`. But with less than three dozen tokens to choose values for, it's better to have coverage for a component that might appear instead of no coverage at all with hundreds of tokens and more added over time.

In later chapters, we'll explore how to manage these new resources and suggest ways to balance responsibility.

Rethinking Relationships

As another exercise, it is common for folks to design an experience where the size of the text is directly related to the space around the text. In fact, this is written within most user agent styles (browser defaults) in a page using CSS. The heading elements use em units to apply the vertical margin. The em unit relates to the current font size, and the more important the heading, the larger the font size. This results in larger margin for larger headings. In this view, it seems natural to consider that the space and font size depend on each other in some way.

This also makes sense when we look at common UI patterns. Large hero sections introducing a visitor to a website often have huge text and spacious layouts, while the dashboards and data tables will often be dense with smaller text to cram as much information as possible. These are design decisions based on the expected needs of the user. And that's the point.

Instead of thinking that the amount of space is dependent on the font size, consider that both the space and the font size are dependent on the mode. In this way, a change applied by the mode would affect both the space and the font size separately, but in the shared context. As illustrated earlier, a hero section could be a mode that is applied due to a design decision: this area should be roomy and inviting. Meanwhile, we could use the same content lockups with a different mode meant to feel more compressed, reducing font size and space to achieve the desired purpose. This makes font size and space not symbiotic, but dependent on the higher context of the mode.

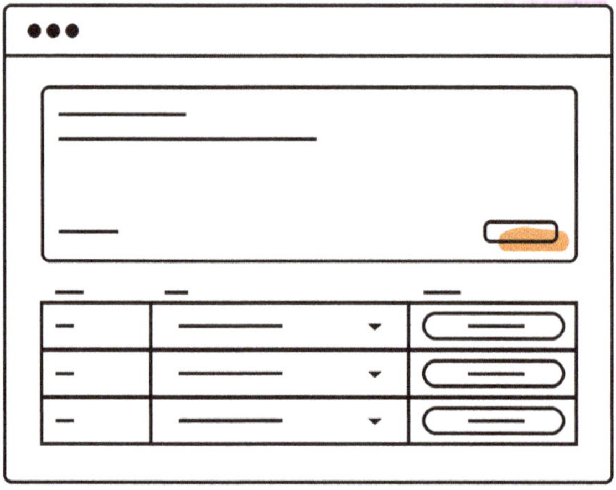

*Let the text inside the elements be determined
implicitly by the surrounding scope*

This results in something that many systems don't provide: *unlimited text sizes*. At first, this sounds like a nightmare from a consistency standpoint. However, the caveat is that the person looking to use a new size doesn't explicitly choose the size they want to use. They would choose a mode that would influence the size in an expected way. Modes could compound to affect size in varying predetermined amounts. We'll see what this might look like in practice later.

If you were considering combining both approaches, such that the font size directly affects the space but both are ultimately affected by a value provided by the mode, consider text priority. The concept of priority was introduced in the discussion about intent categories, where primary text is meant to describe the most important content a person should be reading in a lockup of text content. In a content lockup, we'd expect all priorities of text to be represented in many cases: a headline, body copy, and some optional help text. While each of these will have different font sizes to convey hierarchy, it is unlikely that each has their own amount of space. It is more likely that we want to show the close relationship between this lockup content and therefore prepare an equal amount of space around and between the items.

If you disagree with keeping the relationships uniform, feel free to experiment with using the font size to help influence the related space or to include other contextual factors like device dimensions or input mechanisms.

Expressive Enhancement

A benefit to Mise en Mode is the confidence of coverage. Because each collection of tokens distributed as a mode is meant to be completely curated with an identical set of intents and values, there is no risk of a missing token in the collection, or a component style that doesn't have an appropriate token assigned yet. Based on earlier exercises in intent naming that are meant to cover all user interface possibilities, there should always be an intent that matches a component property. That complete list is your mode, and the values assigned convey the purpose of the mode through presentation.

Let's return to the inverse token approach and when we identify a new inverse token would be needed. If we were to continue down this unlucky path, consider the timeline of events that would occur. For the purpose of this exercise, when I say "we", I'm referring to a design systems team responsible for tokens and shared components.

1. Someone decides the input component needs to be added to the dark-colored footer.
2. We review the current component and all of the tokens that are currently covering style properties.
3. We duplicate all of those tokens specifically for when the input is meant to be inverse.
4. We assign new values to all of those tokens.
5. We publish those changes to the library for teams to use.
6. The team responsible for the footer adds the input from the library using the inverse variation.
7. The site is published with the new input inside the dark-colored footer.

Remember, these steps repeat for every new component needed in the footer. If we suddenly need a pricing table in this section, we'd need to repeat these steps for that, too. And let's hope that a team doesn't try to add the component before these steps. If they do, they'd either be using a component that doesn't yet support existing on a dark background or they'd begin overriding all the styles themselves so that it does. Both directions are detrimental to the platform but the steps to introduce new tokens and styles are bottlenecked so it might look more efficient to DIY.

Now, let's look at the same process using Mise en Mode. Importantly, this will assume that because the dark-colored footer exists as a mode, all the values for the intents that describe the concept of dark have already been curated.

1. Someone decides the input component needs to be added to the dark-colored footer.
2. The team responsible for the footer adds the input from the library.
3. The site is published with the new input inside the dark-colored footer.

While it is obvious that there's a significant reduction in the number of steps, what is more important is that no steps introduced the word "we". In other words, *the design systems team does nothing and everything looks as intended.* The intents assigned within the input component are designed so that they accept any value provided by any appropriately curated mode. The value that is finally applied depends primarily on where the component was placed. In light-colored areas, the input is styled with this mode in mind. In dark-colored areas, an entirely different treatment can be provided in the same place. This is all done separately from interface composition. Pages can be built in parallel to new modes being introduced.

Now let's consider what might happen if the activation of the mode in the dark-colored footer was removed. For example, let's say we forgot to apply the mode to this area. When this happens, the ancestor mode on the page is applied. In the case of the light page with a footer without its own mode, the footer would then adopt the values set in the light mode and present a light-colored footer matching the rest of the page. This is why we use the term **expressively enhanced**. Every mode is meant to enhance the presentation of the underlying structure. When a mode is removed, it'll refer to a parent mode for values. If the references go all the way to the top without values, we would effectively see the bare structure of the experience. This behavior is a benefit because we can be sure that if a mode is not applied, the parent mode will support this area cohesively. A parent mode caring for any missing children.

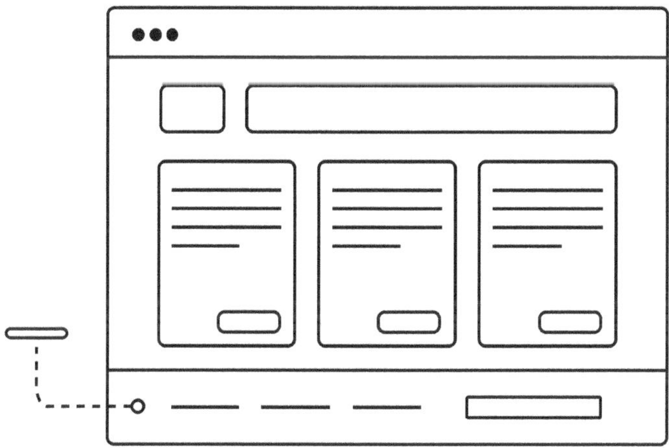

*A light-colored page with a light-colored
footer, marked where the mode is missing*

In the inverse tokens example, any missing part to the steps outlined could result in an unusable experience because the area where the input exists was not fully prepared to support it; having missing tokens or values. The input was designed in isolation, and the footer and earlier ancestor containers have no responsibility in providing values to it. This could cause the input to not receive appropriate values and make the presentation look incomplete, or worse, outright inaccessible.

*A light-colored page with a dark-colored footer,
marked where the inverse token is missing*

This shows that the amount of coordination to support inverse tokens causes unnecessary bottlenecks, slowing iterative development by including too many cooks for a single color choice. Mise en Mode simplifies the process, only including the folks who are responsible for the change. People who care about color can change the color, and in some more advanced systems, can even influence where their new expression is used without bothering other departments.

Empowering Marketers

Here's another scenario where Mise en Mode is a powerful approach. Have you ever had a request from the marketing team for a colored background or button that just isn't supported in the system? Earlier we described how many steps it would take to add this token, and because this request isn't coming from a user need, it feels even less warranted. Marketers want to "make it pop" for a few weeks and we want to keep it consistent for the future. Adding a token that will only exist for a limited run doesn't seem worth our time. However, the case is made that this background color change will drive sales through the roof, so we need a solution.

Imagine that we had a "make it pop" mode. That mode could be applied to places that our marketers want to control to attract focus for pushing features and products. We could even have specific modes for individual features. In a robust system, marketers could have the ability to make changes to these promotional modes and connect them to experiments that collect data and inform marketing strategies. That premium pricing tier meant to draw all the attention would be the responsibility of a marketer, not just choosing the design treatment, but also executing it independently. Wherever the promotional mode is applied could be a separate team's responsibility, so that the marketers don't go too wild with their imagination.

Pricing page with 3 tiers of increasing visual treatment

For some, this could sound like a nightmare. You'd be allowing folks with a historically limited design acumen to be in charge of selecting colors. It's true, this is easy to do poorly. Even for seasoned designers, choosing the right colors can be a challenge.

As mentioned earlier, design systems' maintainers don't typically have opinions about colors past accessible contrast pairs. Transferring this responsibility should be a goal, to free up time for more important work and to hold the people who have opinions about color accountable for those opinions.

Modes in Real Life

At this point, you should see a huge benefit with using modes to expressively enhance areas of our experiences for any purpose a stakeholder might want. However, you may still be apprehensive about starting this journey yourself or with your team. Curating all the design token values for light mode and all the same design tokens for dark mode seems much more labor-intensive than only adding a few inverse tokens for a specific component variation. This is further compounded by the additional presentation expressions you may want to consider, such as delivering the concept of "critical" as a mode instead of a component variant. Each collection of values delivered as a mode will require significant work to curate. Community peer research I conducted shows that it takes nearly two hours on average to curate the value for a single token for a design system. Now imagine multiplying this duration by the number of tokens for each mode you may want to support. If concepts like "critical" are considered modes, you could assume that this exercise could take a few months to cover all expressions fully because the exercise of choosing a value for any particular existing token can be challenging.

From a maintenance perspective, it's completely reasonable to feel overwhelmed with what is being introduced here. Each new mode requires an exercise of curating several dozen tokens when you may have previously felt more comfortable simply introducing a few inverse tokens and getting the intended result with what is seemingly much less to maintain. I believe choosing to do that suggests that your system's needs will never grow or that you understand the consequences and have planned the resources to improve over time. Either way, as outlined earlier, the inverse token approach is technical debt as soon as it is introduced. Instead, I recommend starting with a single mode and adding what you need over time. Due to the expressive enhancement nature of the approach, you can prepare dark mode separately from the new features inevitably being launched as resources permit.

Gaining confidence in this approach takes a lot of time and understanding of use across the practice of user experience design. The earlier exercise—identifying purpose within our experiences—certainly helps, but this knowledge isn't often shared with clear comprehension amongst a larger team. In other words, a person coming across the set of intents will most likely have questions about where `$surface_primary_backgroundColor` comes from and have their own opinions about a more intuitive token. Controversially, the goal here is not for people to feel comfortable with the token naming scheme. **The goal is to create a set of tokens that are disassociated from opinions and are rooted squarely in universally identified user experience patterns that meet user expectations.** In the ideal world, once an intent is assigned to a component, we should never need to reconsider its application within that component. In a mature component library, very few people should be in the weeds of intent assignment in components, as newer features should be composed of smaller parts of the component library. Complete experiences should need very few custom assets and intent assignments to present a final product.

Another pill that is hard to swallow: the separation of responsibility. In Mise en Mode, decisions such as what color the button is are assigned to the owners of a mode, not to the folks composing an interface. Remember, this approach is structure-first. So product designers may feel like their former opinions about presentation responsibilities are now reassigned to others. This is another opportunity to make the curation a shared responsibility. Any designer who has an opinion about what color to use should be able to make a case for it, knowing that the choice will cascade throughout the platform. You could imagine an experimentation platform wired to the modes, where a new experimental expression could be tested and data could be collected. The decision could also be peer-driven, where the change is determined by a vote. No matter the process, bringing the folks who have opinions about a mode's values (i.e., expressions) will offload even more responsibility away from the design systems team. The best case is for the organization at large to be responsible for curating the values of modes. The design systems team merely facilitates decisions, leaving others free to create whatever expressions their hearts desire.

Importantly, the Mise en Mode approach allows for any combination of values in a complete expression to be applied anywhere in an experience. In a theoretical example, every stakeholder at a company could curate their own mode. Even the CEO could have their own mode if they wanted! That's a huge benefit to the approach; supporting highly personalized changes to the interface with very little involvement from other members of the organization. In just a few hours of curating values, a new expression is born.

If this is scary to you, it suggests you are expecting to control these expressions in some way. Perhaps you're worried that the mode they create isn't accessible or conveys the wrong message. Instead, think of this as empowerment and transferring responsibility to open up alternative work for yourself and your team. Deep down, you know the color of the button doesn't really matter more than a user trusting the brand. Let that decision go into the hands of someone who wants that responsibility.

A New Responsibility

So here's the important new role for a designer in this world. Instead of going through an exercise in token naming, and in some cases choosing values, the exercise is now *curating where this expression will be used in our experiences*. A designer would identify areas of the experience where a change of expression should occur and mark them to be expressively enhanced using a mode that describes the expectation. Each new expressive mode could either be used repeatedly across the experience or only once for a unique mood.

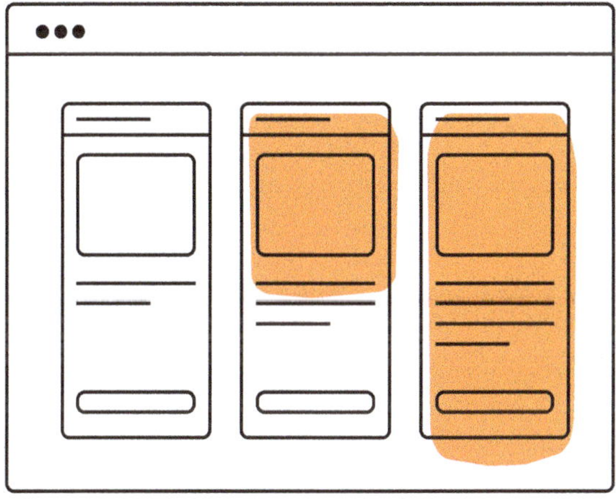

*Choose to apply an expression to
part of a region or an entire area*

This approach introduces a separation of design responsibility. What those new values are for any mode may not be up to the same designer curating the placement of those modes. This is a difference between UX and UI. In the UX mindset, we want to determine the most optimal path and composition of elements for the user and make a decision that works towards successfully meeting their own needs. Meanwhile, we present these experiences through presentational UI design decisions such as color, typography, etc. For larger organizations, we often see specific designers responsible for product features (UX) while others are responsible for the product's look (UI). It is not uncommon for these designers to blend responsibilities or have influence in both areas. However, this separation allows for a better focus of responsibility for how a product should be presented to the user. It's the difference in quality that is often found between a general contractor and several specialists.

Chapter Summary

Modes represent a fundamental shift from creating duplicate tokens for every variation to curating complete collections of token values that can be applied to any scope. Rather than building inverse tokens for dark footers or special tokens for critical states, we apply modes—fully curated sets of values for all existing tokens—to specific boundaries within an experience. Mise en Mode means elements receive new values based on where they exist, not through new tokens. A dark footer simply uses dark mode within its scope; a critical alert activates a critical mode that transforms everything within its boundary. Modes can nest within modes, enabling unlimited expressive possibilities without exponentially growing your token library.

This structure-first approach provides important advantages: it eliminates bottlenecks since new components automatically work in any mode, ensures complete coverage through fully curated collections, and enables graceful degradation where missing modes inherit from parent contexts. Most importantly, it separates responsibilities. UX designers curate where modes apply to serve user needs, while stakeholders who care about specific expressions (marketers, brand designers, feature teams) independently control the values within their modes. Though the transition requires patience and careful planning, starting with a single mode and building coverage gradually creates a scalable system that empowers stakeholders, reduces maintenance burden, and supports diverse expressions without the technical debt of traditional token approaches.

Benefits of Mise en Mode

- Reduces token proliferation
- Separates concerns between structure and expression
- Enables scalable design variations
- Empowers different stakeholders

Key Concepts

- Start with a single mode
- Build mode collections gradually
- Focus on complete coverage
- Consider organizational impact

Best Practices

- Use modes for systematic variations
- Maintain clear scope boundaries
- Document mode relationships
- Plan for mode inheritance

File Management

At this point, you should have grasped the core concepts of Mise en Mode. A small collection of design tokens can have their values changed based on a new scope within the experience. It's time to make the dream a reality and create some infrastructure to support the concept.

The next few chapters are going to have a heavier reliance on demonstrating code. If you're uncomfortable with code, you should still continue reading these sections. We'll be covering additional considerations in the overall system that can influence additional design needs. For example, how outside factors could influence a mode or how to determine which modes to deliver first. If you're working with a developer, you should be able to collaborate with them to make the best decisions based on these factors and have some understanding behind decisions made in this architecture. Code will be reviewed in following paragraphs for further comprehension.

If you're comfortable with code and want to follow along, here's a minimal project structure that we'll be using. We'll be covering each file over the course of the next few chapters. In some cases, we may revisit files or suggest where enhancements could be introduced.

```
├── /components
│   └── _tokens.module.scss
├── /infrastructure
│   ├── bootstrap-mode.js
│   ├── get-interop.js
│   ├── get-inventory.js
│   ├── get-schema.js
│   ├── intents.yml
│   ├── paths.js
│   ├── to-list.js
│   ├── to-var.js
│   └── write-files.js
├── /modes
│   └── _schema.json
└── /public
    └── _inventory.json
```

There's also a complete GitHub repository[1] with all of the finished code to help follow along.

The files prefixed with an underscore (_) character are generated from other files. I'll be providing the code as vanilla JavaScript. We'll also be using some lesser known techniques to help manage outputs. Feel free to adjust the code where appropriate to your capabilities.

Before we continue, I'll provide some of the content for a few of the files. It will help to have these in place before we get to some of the larger logic later on.

```js
import { join } from 'node:path';

// Components, used to demonstrate interoperable
tokens.
export const COMPONENTS_DIR = join(process.cwd(),
'components');

// Infrastructure, holds most of the scripts
writing files.
export const INFRASTRUCTURE_DIR =
join(process.cwd(), 'infrastructure');

// Modes, place to modify expressive modes as YAML.
export const MODES_DIR = join(process.cwd(),
'modes');

// Public, hosts files (eg., CSS) to the web page.
export const PUBLIC_DIR = join(process.cwd(),
'public');

// The filepath to list of all intents.
export const INTENTS_YAML_PATH =
join(INFRASTRUCTURE_DIR, 'intents.yml');

// The default filepath created during mode
bootstrapping.
export const NEWMODE_YAML_PATH = join(MODES_DIR,
'newmode.yml');

// The filepath for the mode YAML schema.
export const SCHEMA_JSON_PATH = join(MODES_DIR,
'_schema.json');

// The filepath for interoperable tokens.
export const TOKENS_SCSS_PATH =
join(COMPONENTS_DIR, '_tokens.module.scss');

// The filepath to host the inventory for the mode
manager.
export const INVENTORY_JSON_PATH = join(PUBLIC_DIR,
'_inventory.json');
```

This is the contents of the `infrastructure/paths.js` file which
helps provide a reusable construction of directory and filepaths

for the project. These are references to the files that we'll need to be read and written; creating further resources for the system to use.

> **ⓘ NOTE**
>
> Note that this uses `process.cwd()` to determine the current working directory where the script was run. It's possible that you may need to update the paths for your own project. Luckily, they can also be updated in one place, within the file.

infrastructure/to-list.js

```js
export function toList(items, fn) {
  return items.map(fn).join('\n');
}
```

This is the contents of the `infrastructure/to-list.js` file. It is a single function, but the logic gets reused in a few places within the infrastructure to help build CSS and CSS-like files. If you find this to be too specific of an abstraction, you can certainly use the `Array.map().join()` in place instead.

infrastructure/to-var.js

```js
export function toVar(intent) {
  return intent.replace('$', '--🔒');
}
```

This is the contents of the `infrastructure/to-var.js` file. While this also looks oddly specific, I'd highly recommend to keep this reusable as the logic here helps write the CSS meant to deliver the mode values in a standardized way. We'll discuss more about what this supports later in this section.

Human Maintainable

Traditionally, you may have a single file to handle tokens or perhaps several files specifically to manage primitives like color and typography. With the introduction of modes, you would be expected to handle many more files—one for each mode. This section aims to describe some best practices to make handling multiple modes as easy as possible and prepare for changes over time.

A core principle that I hold when managing configurations is for a human to edit the file easily. Historically, our practice has used JSON to describe tokens. This format has been helpful since it is understandable by many computer systems. This is also its drawback—it is primarily meant for computer systems, not humans. Instead, I recommend authoring in YAML. These files can be converted to JSON through some small processing that could be included in any toolchain. Let's look at the comparison between these in the following examples, first here is some data as JSON:

```json
{
  "$some_token": {
    "$value": "red"
  }
}
```

This is the identically structured data as YAML:

```yaml
$some_token:
  $value: red
```

I'll argue that the YAML is objectively easier to read and write due to the lack of additional characters to describe the tree-like structure. YAML uses whitespace to show nesting. The number of space characters indicates how a value is nested under another value. It also has the added benefit of comments, something that JSON itself cannot provide.

There are extensions of JSON that do allow comments among other improvements. However, the additional characters are still required to show the nesting.

Speaking of comments, YAML comments start with the # symbol. So if you're using hexcode colors, you'll need to wrap the color in some quotes so it doesn't get ignored as a comment in processing.

To start this, we want to have a list of tokens that informs the rest of the system. We prepared this list earlier in the book, but this is where we'll finally write it all down. All the other files for the system will be based from this list. Our prototypical list of tokens should look like this:

```yaml
# Form controls (3)
- '$control_backgroundColor'
- '$control_foregroundColor'
- '$control_borderColor'

# Buttons and links (9)
- '$action_primary_backgroundColor'
- '$action_primary_foregroundColor'
- '$action_primary_borderColor'
- '$action_secondary_backgroundColor'
- '$action_secondary_foregroundColor'
- '$action_secondary_borderColor'
- '$action_auxiliary_backgroundColor'
- '$action_auxiliary_foregroundColor'
- '$action_auxiliary_borderColor'

# Non-interactive containers (9)
- '$surface_primary_backgroundColor'
- '$surface_primary_foregroundColor'
- '$surface_primary_borderColor'
- '$surface_secondary_backgroundColor'
- '$surface_secondary_foregroundColor'
- '$surface_secondary_borderColor'
- '$surface_auxiliary_backgroundColor'
- '$surface_auxiliary_foregroundColor'
- '$surface_auxiliary_borderColor'

# Font metrics (12)
- '$text_primary_fontSize'
- '$text_primary_fontWeight'
- '$text_primary_fontFamily'
- '$text_primary_lineHeight'
- '$text_secondary_fontSize'
- '$text_secondary_fontWeight'
- '$text_secondary_fontFamily'
- '$text_secondary_lineHeight'
- '$text_auxiliary_fontSize'
- '$text_auxiliary_fontWeight'
- '$text_auxiliary_fontFamily'
- '$text_auxiliary_lineHeight'
```

This is the data saved in the `infrastructure/intents.yml` file. If
this file was to be transformed into JSON, the result would be an
array of strings, each being a token name. We'll be using the

package `js-yaml` to execute this transformation in a few places. Be sure to include it in your project. This will be the only outside dependency we'll need.

If you're writing YAML in a code editor, you should look into adding an extension that helps with YAML syntax and validation while authoring `.yml` files. The extension will depend on your particular editor of choice. After activating, check for new helpful highlighting in the `intents.yml` file.

Mode Files

Now that we have this list of tokens, we'll also be using YAML to write the curated values to our tokens, what we might normally call a theme. I'll be referring to these as "mode files" since the purpose of us doing this work is to manage modes and each theme is meant to express a mode. Here's an example of our primary surface tokens with some basic colors curated in a `.yml` file:

modes/sample.yml

```
mode: sample
tokens:
  $surface_primary_backgroundColor:
    $value: white
  $surface_primary_foregroundColor:
    $value: black
  $surface_primary_borderColor:
    $value: transparent
```

This file should be saved as `modes/sample.yml`. Notice that we've also included the identifier of the mode in the file. The `mode` should be the identifier that applies the mode to an area of the page. You can think of this as `light`, `dark`, `promo` or any other name that best describes *why* this set of values exists; its purpose. This will be useful for us later. The `tokens` key holds the collection of token-value pairs curated by a stakeholder. I recommend using the word `tokens` here over intents as this is a more common term and folks outside of the system may be responsible for curating these files.

You might also notice that I've opted to keep the file structure relatively flat. This is in comparison to other token management systems which provide a deep nesting structure. A nested structure requires a lot of recursive tree traversal to finally determine a token's value. The flatter nature of this will make token manipulation easier later in our work.

Our final file will be larger, handling the few dozen tokens we've discussed earlier. This resulting file should only be around 100 lines long, compared to several hundred that formatted JSON would require with its additional markup.

You might be wondering why there's an extra key called `$value` at the end of each structure instead of simply setting the value at the end. This will help include additional metadata about the token. Even if you don't have metadata for these tokens now, you'll be thankful that you are prepared for this if and when you need it later. Importantly, folks using tokens most likely will not see these metadata keys. So I wouldn't worry about the $ prefix being confused as a token.

> ⓘ **NOTE**
>
> I use the $ symbol to denote special keys that are expected within the token structure. It is possible that teams working in these files might want to include additional metadata here. The $ can indicate which keys are necessary for the system to function properly from ones that are more optional or not as important. Feel free to use your own convention if this is confusing between their use here and in token names themselves.

One example where I find metadata helpful is specifically for the `fontFamily` entry since we can supply the branded font as well as a fallback.

modes/sample.yml

```
mode: sample
tokens:
  $text_primary_fontFamily:
    $value: Plus Jakarta Sans
    $fallback: sans-serif
```

This works well to separate fonts we expect to include additional resources for and ones that are considered "web-safe". In other words, fonts that are most likely preinstalled on users' computers which we can reliably fallback on if our custom font doesn't load properly. The $ prefix indicates this key will be used to help compose the value later in the lifecycle.

To help humans that need to create new modes, you'll want to consider including some script that generates a starter template file to get a person started with all the tokens listed awaiting values. Providing this will also help ensure that every mode is filled out in its entirety for full expression coverage. We'll address how to provide this after introducing a YAML validation schema.

Human Errors

The drawback of preparing a file to be ready for human editing is that humans will edit the file. Humans are prone to errors and one wrongly typed token can produce some poor results. Luckily, we have ways of guiding humans so fewer mistakes are made.

The goal will be to create a schema for our YAML token files to follow. A schema for YAML is created as a JSON file. Assuming you have a YAML file that begins like this:

modes/sample.yml

```
mode: sample
tokens:
  $surface_primary_backgroundColor:
    $value: Canvas
```

The general structure for your `modes/_schema.json` file will be
generated to eventually look something like this:

modes/_schema.json

```json
{
"$schema": "http://json-schema.org/draft-
07/schema#",
  "type": "object",
  "properties": {
    "mode": {
      "type": "string"
    },
    "tokens": {
      "type": "object",
      "properties": {
        "$surface_primary_backgroundColor": {
          "type": "object",
          "properties": {
            "$value": { "type": ["number",
"string"] }
          }
        }
      }
    }
  }
}
```

The underscore character in file name `_schema.json` indicates
that this file is generated. It should not be edited by hand. We'll
get to how to do this in a moment. This file should exist in the
same place as your mode files, in the `modes` folder. This should
be the only `.json` file in this folder.

For each expected key in your YAML structure, we need to
define `"type"` and `"properties"`. The `"type"` will almost always
be `"object"` until defining the final `"$value"`. When you finally

define the "$value" at the end, this will most likely include the "type" as either "string" or "number". You can also define multiple types using an array syntax (e.g., ["string", "number"]). There are additional items you can add to the validation that can introduce required keys and messaging. We'll skip that to have something to start with.

As mentioned, you don't want to hand write this file yourself, you'll want to generate it using our expected tokens structure. For most of the management, we'll want to have assets generated wherever possible. Below are the functions we'd expect in the infrastructure/get-schema.js.

infrastructure/get-schema.js

```js
import { readFileSync as read } from 'node:fs';
import { load } from 'js-yaml';

import { INTENTS_YAML_PATH } from './paths.js';
```

This is the top of the file, where we import some functions to help soon read the contents of the infrastructure/intents.yml and load as a JavaScript object.

infrastructure/get-schema.js

```js
function createValue(intent) {
  const base = {
    $value: {
      type: ['number', 'string']
    }
  };

  return base;
}
```

The createValue function is responsible for creating the final nodes within the schema. You might notice that we're passing in the intent here, but not currently using it within the function. This is where you can introduce additional validation, such as checking the intent name for fontFamily and adding a

`$fallback` key to the schema. At least, we'll want to ensure that
this node has a `$value` and we're currently allowing that to be
either a `number` or `string`.

infrastructure/get-schema.js

```javascript
function createTree(intents) {
  return intents.reduce((acc, intent) => {
    return Object.assign(acc, {
      [intent]: {
        type: 'object',
        additionalProperties: false,
        required: ['$value'],
        properties: createValue(intent)
      }
    });
  }, {});
}
```

The `createTree` function is the main iterator which goes
through the list of intents to create all of the nodes for the
schema. If you aren't familiar with options within the schema,
here's some quick explanations:

- The `type` key tells the validator what kind of value is
  expected on this key. Since we're going to be adding an
  object to hold `$value`, along with the possibility of other
  metadata, this is an `object`.
- The `additionalProperties` effectively freezes the
  schema at this node. This means no further properties are
  allowed past what is defined at the `createValue` function.
- The `required` key shows what must be set at this node. As
  we might expect, we want to ensure each intent has a
  `$value` set.
- The `properties` key defines all of the properties we
  expect on the object. This is written from our earlier
  `createValue` function.

This demonstrates one of the main reasons for having a flat
token structure instead of a nested one. A flat structure allows

for a simple way of iterating over all of your tokens for
processing. If the tokens were a more nested structure, looping
through all of the tokens becomes more cumbersome.

infrastructure/get-schema.js

```js
function createSchema(intents) {
  return {
    $schema: 'http://json-schema.org/draft-07/schema#',
    type: 'object',
    required: ['mode', 'tokens'],
    properties: {
      mode: {
        type: 'string',
        additionalProperties: false
      },
      tokens: {
        type: 'object',
        additionalProperties: false,
        properties: createTree(intents)
      }
    }
  };
}
```

The `createSchema` function handles the main schema body,
setting up the initial nodes. You should be familiar with most of
the keys since they have the same purposes from the earlier
function. The `$schema` key is a meta-reference, explaining how
this schema is meant to be generally constructed.

infrastructure/get-schema.js

```js
export function getSchema() {
  const intents = load(read(INTENTS_YAML_PATH, 'utf8'));
  return createSchema(intents);
}
```

This is the final function of the file, `getSchema`, which is the only
export here. This is meant to read and load the

`infrastructure/intents.yml` and send it into the `createSchema` function to generate the JavaScript object that will eventually be written as JSON.

> (i) **NOTE**
>
> I'd recommend keeping the schema as an object instead of immediately writing the `_schema.json` here. We'll be using this data to help generate the template that creates new mode YAML.

Let's start adding code to the `infrastructure/write-files.js` file to create the `modes/_schema.json`.

infrastructure/write-files.js

```js
import { writeFileSync as write } from 'node:fs';
import { getSchema } from './get-schema.js';

import {
  SCHEMA_JSON_PATH
} from './paths.js';

const _schema = getSchema();

write(
  SCHEMA_JSON_PATH,
  JSON.stringify(_schema, null, 2),
  'utf8'
);
```

There shouldn't be much in this file to start. We'll be including more over time to create a few other files. At this point you should be able to run `node ./infrastructure/write-files.js` in a terminal for the project and see the `modes/_schema.json` appear.

Making Modes

Now we're ready to start creating a mode file. As mentioned earlier, we want to give people a starting point to begin curating

and not just a blank file. We can generate a template based on
the schema. This is the main reason why we shouldn't
immediately write a JSON string from the `getSchema` function.
We can use it to help create a YAML template:

infrastructure/bootstrap-mode.js

```js
import { writeFileSync as write } from 'node:fs';
import { dump } from 'js-yaml';
import { getSchema } from './get-schema.js';

import { NEWMODE_YAML_PATH } from './paths.js';

// Used for YAML validators to refer to schema.
const COMMENT_HEADER = '# yaml-language-server:
$schema=./_schema.json';
```

This is the header of the `infrastructure/bootstrap-mode.js`
file. It imports our previously made `getSchema` function, along
with some functions to help write the resulting YAML. The
`COMMENT_HEADER` is the secret sauce to connecting the mode file
with the validation schema from earlier. Every mode file created
should include that comment to trigger YAML validation and
help with the authoring process. Here's what that comment
should look like in the YAML file:

```yaml
# yaml-language-server: $schema=./_schema.json
```

This assumes that the `_schema.json` file is in the same directory
as the current mode file. If everything is prepared correctly, any
time the mode file doesn't have the correct structure, you
should see an indicator in that file and messaging explaining the
issue. We'll eventually be writing this YAML comment into our
bootstrapped mode file.

infrastructure/bootstrap-mode.js

```
function merge(acc, [property, value]) {
  return Object.assign(acc, { [property]:
recurse(value?.properties) });
}

function recurse(properties) {
  if (!properties) return null;
  return Object.entries(properties).reduce(merge,
{});
}
```

These two functions, `merge` and `recurse`, are reading the complex nested schema tree we created earlier. As we visit nodes, we determine if another object should be created or if we should end with a `null` placeholder. That `null` placeholder will be invalid according to our schema, highlighting that these values need to be addressed.

infrastructure/bootstrap-mode.js

```
const _schema = getSchema();

const data = recurse(_schema.properties);

write(
  NEWMODE_YAML_PATH,
  [COMMENT_HEADER, dump(data)].join('\n'),
  'utf8'
);
```

For the last part of the `infrastructure/bootstrap-mode.js` file, we generate the schema, then we recurse over the properties to create a new object that represents the expected structure of a mode based on the `modes/_schema.json` file. Finally, we write (or, using the term from `js-yaml`, "dump") that object as YAML into a new `newmode.yml` file. If you run `node ./infrastructure/bootstrap-mode.js` in the project root, you should see the new file under the `/modes` directory, filled with `null` values as placeholders for the expected intents.

Setting up the project in this way will help if and when new tokens are added to the system. Once the `intents.yml` file is updated and scripts are re-run, the `_schema.json` will also update and show the new tokens in newly created mode files.

 **WARNING**

With what we have so far, updating the `intents.yml` will *not* cause warnings in existing mode files for missing token entries. You *could* update the schema to make these tokens required. However, since we may want to split mode files into collections of similar properties, it is possible that a mode file will not be expected to have complete coverage itself.

You could include additional metadata marking a particular file to expect color or typography only and then mark the necessary tokens as required. When tokens are marked as required in the schema, you'll be notified in the file when they don't exist.

Modes as CSS

Next, we'll want to read each file in `modes` and convert it into a `.css` file of CSS Custom Properties, also known as CSS variables. In the lifecycle of the application, we'll be requesting these files to convey new expressions. You'll want to prepare for these files to be publicly served as a part of the user experience, since a mode can be requested *after* the application has finished loading. We'll discuss more about the lifecycles in which a mode can be delivered later. For now, here's the most simple version of a function that will prepare CSS variables from a mode file, starting with the file header:

infrastructure/get-inventory.js

```js
import { readFileSync as read, readdirSync } from
'node:fs';
import { join, parse } from 'node:path';
import { load } from 'js-yaml';
import { toList } from './to-list.js';
import { toVar } from './to-var.js';

import { MODES_DIR } from './paths.js';
```

This is the beginning of the `infrastructure/get-inventory.js`
file. You may have expected this file name to be more
descriptive about writing CSS. While that will be a large purpose
of this script, it will not be the main purpose. We'll uncover the
importance of the inventory soon.

infrastructure/get-inventory.js

```js
function toDeclaration([intent, metadata]) {
  const decl = [toVar(intent), metadata.$value];
  return `${decl.join(':')};`;
}

function toCss(mode, tokens) {
  const declarations =
toList(Object.entries(tokens), toDeclaration);
  return `[data-mode~="${mode}"] { ${declarations}
}`;
}
```

These are the two functions responsible for generating the
required CSS from the YAML found in the `/modes` directory. The
result of these functions will eventually be CSS that looks like
this:

public/_sample.css

```css
[data-mode~="sample"] {
  --🔒surface_primary_backgroundColor: Canvas;
  --🔒text_primary_fontFamily: Roboto;
}
```

The `toVar` function we set up earlier handles the conversion of an `intent` name to a CSS Custom Property; replacing $ for -- 🔐. There are a few reasons for this:

1. CSS variables *must* start with a -- to be valid and usable. Additionally, the $ character is not a valid character to include in this CSS construction but emojis are.
2. The use of this emoji makes it hard to simply retype this variable by hand. This suggests that there's some other way to get this resource other than typing, such as copy and pasting or by import reference.
3. An emoji can convey more information in one character of space than alphanumeric characters can.
4. When reviewing the output of this in a browser's developer tools, the resolved CSS variables are much easier to identify since the emoji stands out from other CSS variables without the emoji. This makes them more special than others that might be used in the same declarations.

The idea behind this emoji is for the variable to be private. Common convention for this in other languages that don't have the concept of private variables is the _ character. However, this allows the variable to be typed easily and may be confused for an incomplete token due to the further use of _ as a delimiter elsewhere in the name. If you want to choose another emoji for this purpose, be sure that it isn't confused for different messaging. The 🚫 could be misunderstood as "deprecated" or "broken". Perhaps the 🪙 could denote a token. This is certainly an area where some personality could be included. Maybe even referencing your design system's personality directly.

This is also the place where you may consider handling more complex mode data structures. For example, consider the following:

infrastructure/get-schema.js

```js
function createValue(intent) {
  const base = {
    $value: {
      type: ['number', 'string']
    }
  };

  if (intent.endsWith('fontFamily')) {
    base.$fallback = {
      type: ['string', 'array'],
      items: { 'type': 'string' }
    };
  }

  return base;
}
```

Here, we're defining a new allowed key in the YAML schema that can have font fallbacks defined, either as a string or a list of strings. Then a person can supply values for the font fallbacks:

modes/sample.yml

```yaml
mode: sample
tokens:
  $text_primary_fontFamily:
    $value: Plus Jakarta Sans
    $fallback:
      - Helvetica
      - Arial
      - sans-serif
```

To handle the `$fallback` values for `fontFamily`, you'd want to catch the usage within the `toDeclaration` function and write the appropriate syntax for defining `font-family` fallbacks:

```js
function toDeclaration([intent, metadata]) {
  const decl = [toVar(intent), metadata.$value];

  if (intent.endsWith('fontFamily')) {
    decl[1] = [decl[1]]
      .concat(metadata?.$fallback)
      .filter(Boolean)
      .join(',');
  }

  return `${decl.join(':')};`;
}
```

> ⓘ **NOTE**
>
> This logic does not account for loading web-fonts. You
> might want to include some additional handling that looks
> for web-fonts and write the appropriate `@font-face`
> declarations. How you write the `@font-face` declaration will
> be dependent on how you lookup and serve your fonts. For
> example, perhaps you'd want to include `size-adjust` and
> `ascent-override` CSS properties to help with layout
> shifting when a new mode is loaded. Font handling is just
> outside the scope of what we're discussing for the
> purposes of this work.

infrastructure/get-inventory.js

```js
export function getInventory() {
  const files = readdirSync(MODES_DIR);

  const inventory = files.reduce((acc, file) => {

    // We'll be adding lines of code in here...

  }, []);

  return inventory;
}
```

This is the main export of our `infrastructure/get-inventory.js` file. A loop that reads the files within the `modes` and will eventually generate the `.css` files, but primarily creates an `inventory` of the modes. We'll be using this inventory to look up data about the modes that a page might need. But since we're also running through all of the modes to do this, we might as well generate the CSS while we're here.

Here's the lines of code we'll need within the file reducer:

infrastructure/get-inventory.js

```
const { name, ext } = parse(file);

if (!ext.endsWith('yml')) return acc;

const filepath = join(MODES_DIR, file);

const {
  mode,
  tokens,
  ...metadata
} = load(read(filepath, 'utf8'));

if (!mode) return acc;

const css = toCss(mode, tokens);

return acc.concat({
  mode,
  css,
  href: `_${name}.css`,
  ...metadata
});
```

There's a lot happening in this loop, which is why we broke this out into its own section to review. The first step is to parse the file name and extension. We'll want to check that the extension we're reading is a `.yml` file, and not our `_schema.json` which also exists in the `/modes` directory.

Next, we construct the file path and then read the YAML file contents as a JavaScript object. If a `mode` hasn't been set, then we will not be able to determine when to load this mode into the page. We generate the CSS with the `toCss` function and finally create a payload for the reducer's accumulator.

 **WARNING**

The construction of the `href` might need to be updated based on your project's configuration. This will be the string written into HTML `<link/>` tags to reference the resulting `.css` files.

The result of the reduction is the `inventory` object which will eventually need to be written as JSON. Importantly, we will not need the CSS that was generated to be included in that JSON data, so we can add a little trick to omit the CSS from the stringified data. Place the following code just before returning the `inventory` from this function:

infrastructure/get-inventory.js

```
inventory.toString = function () {
  return JSON.stringify(
    this.map(({ css, ...rest }) => rest),
    null,
    2
  );
}
```

Now, when you stringify the `inventory` out of this function, it'll return it as JSON without the `css`, keeping the size of the `public/_inventory.json` file smaller than it would otherwise be. This will be important later.

Let's update the `infrastructure/write-files.js` file to also write the files coming from this work:

infrastructure/write-files.js

```js
import { writeFileSync as write } from 'node:fs';
import { join } from 'node:path';
import { getSchema } from './get-schema.js';
import { getInventory } from './get-inventory.js';

import {
  SCHEMA_JSON_PATH,
  PUBLIC_DIR,
  INVENTORY_JSON_PATH
} from './paths.js';
```

This will be the new header of `infrastructure/write-files.js` to include the new `getInventory` import and paths.

infrastructure/write-files.js

```js
const inventory = getInventory();

for (const { href, css } of inventory) {
  write(
    join(PUBLIC_DIR, href),
    css,
    'utf8'
  );
}
```

This will write the `.css` files from the generated `inventory`. Take special care in where these files are written and how they relate to the `href` that will be used for the `<link/>` elements.

infrastructure/write-files.js

```js
write(
  INVENTORY_JSON_PATH,
  String(inventory),
  'utf8'
);
```

Finally, we write the `public/_inventory.json` file using our little trick to forcefully omit the CSS from the `inventory` before it is written. This should result in a file that has contents that look something like this, assuming you have no additional metadata:

public/_inventory.json

```json
[{
  "mode": "light",
  "href": "_light.css"
}]
```

Now that we have all of this set up, we need the resources that folks will use to apply tokens into components.

Interoperable Tokens

The previous section wrote CSS variables that looked like this:

```
-- 🔒 surface_auxiliary_backgroundColor
```

As mentioned, we don't expect other users to try and write these tokens directly. Instead, we want to give others an easier interface to reference these tokens when authoring components.

> ### ⓘ NOTE
>
> Since each project could be set up with a number of different frameworks, libraries, or tools, this section is going to make some recommendations that should apply to most codebases. It is possible you'll need to make some adjustments in order to make it work for your organization.

The goal is to provide resources that are universally accessible in a similar way between writing styles in CSS or in JavaScript. This makes it easy for folks who want to use tokens in either ecosystem to naturally reference the name in a nearly identical way. To do this, we'll need to enhance our CSS processing to allow importing variables in a similar way to how we'd import in JavaScript. Here's an example of what we're after:

```javascript
import tokens from '_tokens.module.scss';

function SomeComponent() {
  const style = { color:
tokens.$surface_auxiliary_foregroundColor };
  return <div style={ style }/>
}
```

In this example, a developer has used JSX to create a component. JSX is a syntax that helps create elements declaratively by writing HTML-like notation in JavaScript files. In this example, we are updating the `style` attribute in the `<div/>` to use the value found at `tokens.$surface_auxiliary_foregroundColor`. We haven't created this file yet but the value found at that path should use the CSS variables we wrote earlier. This means we should eventually see the following when inspecting the element in the browser:

```html
<div style="color: var(-- 🔒
surface_auxiliary_foregroundColor)"></div>
```

You may have noticed the hint in the earlier JavaScript that in order for this to work interoperably, we'll need to import using Sass[3]. Sass is a CSS extension language that adds features that aren't available in native CSS. There's a few flavors of Sass—in the examples we'll be writing, I'll be using SCSS. The difference is in the syntax: SCSS is closer to native CSS syntax. This makes the authoring of Sass in CSS feel more familiar. However, if you are already authoring Sass feel free to continue to do so.

We'll want to create Sass variables that can be referenced in other Sass files. To do that, we will use the `@use` rule and reference that same file.

```scss
@use 'tokens.module';

.some-component {
  color: tokens.$surface_auxiliary_foregroundColor;
}
```

If you compare the syntax between the JavaScript usage earlier with this Sass usage, you should notice that the way we reference the tokens is exactly the same. You should also notice that this matches the same construction as we originally introduced with the `$` and `_` symbols. This uncovers why the name construction is so specific. The `$` symbol is a Sass syntax that identifies a variable. And the `_` characters are a delimiter that can work between Sass and JavaScript files. Other common delimiters would be invalid in CSS (`.`) or invalid in JavaScript (`-`).

You might also notice that the file path in the import statements between the two examples are not the same. The curve ball is the Sass import expects an underscore character (`_`) at the beginning of the file name. Then the name following the underscore is the name used to reference the variables written in the file. So the `_tokens` part allows something like `tokens.$var` to be accessible in other `.scss` files.

You should also notice that the file name includes the word `module`. This is an indication of a requirement to use CSS Modules[4], which is a method of importing CSS into JavaScript files. Typically, this is done by creating hashed class names in a CSS files that are then applied to the elements in JSX. However, we're going to leverage a lesser used feature within CSS Modules called "Interoperable CSS". This adds two additional pseudo-selectors. We're going to focus on the `:export` one. Here's the start of the `_tokens.module.scss` file:

components/_tokens.module.scss

```scss
$surface_auxiliary_foregroundColor: var(--🔓
surface_auxiliary_foregroundColor);

:export {
  #{'$surface_auxiliary_foregroundColor'}:
$surface_auxiliary_foregroundColor;
}
```

There's two main parts to this file. The upper section defines the Sass variables and then assigns the CSS variable as the value. The Sass variables will be available at the `@use` `'tokens.module'` statement in other files. The lower section is creating the JavaScript versions of those variables by referencing the earlier Sass variables. You could rewrite the CSS variables instead, but then you'd also need to perform additional interpolation[5]. We do need the interpolation when writing the keys under the `:export` declaration. This is the `#{}` syntax that allows us to write a string as shown in the example above.

Assuming that we have the necessary resources installed into your project to support Sass and CSS Modules, we should now be able to successfully reference one file for tokens with identical authoring experiences! Depending on the environment you use to code components, we should also begin seeing some autosuggesting behavior to help with the authoring experience too.

Clearly, with the amount of tokens we have, we don't want to author this file by hand. We'll want to generate this file using the resources we have put together already. Let's start creating the `infrastructure/get-interop.js` file:

infrastructure/get-interop.js

```
import { readFileSync as read } from 'node:fs';
import { load } from 'js-yaml';
import { toList } from './to-list.js';
import { toVar } from './to-var.js';

import { INTENTS_YAML_PATH } from './paths.js';
```

The header of the file has some functions to help read and load YAML, specifically our `infrastructure/intents.yml` file, along with a few of our utilities from earlier.

```js
function toCSSVar(intent) {
  return `var(${toVar(intent)})`;
}

function toSCSSVar(intent) {
  return `${intent}: ${toCSSVar(intent)};`;
}
```

These functions help create the SCSS variable assignments for
the export. Wrapping the result of `toVar` in a native CSS `var()`
function resolves the value provided by the incoming mode. We
want this to be the value for the related SCSS variable.


```js
function toInterpolated(intent) {
  return `#{'${intent}'}: ${intent};`;
}

function toModuleExports(intents) {
  const interpolated = toList(intents,
toInterpolated);
  return `:export { ${interpolated} }`;
}
```

These functions are responsible for preparing the variables as a
CSS Modules export. The use of the #{' '} interpolation syntax
tells the SCSS ecosystem to write the string as-is, without
resolving the variable. Meanwhile, we write it directly as the
value to this assignment in the `toInterpolated` function. This is
finally written within a `:export` selector, which is an extension of
the CSS Modules specification.

infrastructure/get-interop.js

```js
export function getInterop() {
  const intents = load(read(INTENTS_YAML_PATH,
'utf8'));
  return [
    toList(intents, toSCSSVar),
    toModuleExports(intents)
  ].join('\n');
}
```

Like many of our other files, the final function is prepared as an
export. This reads and loads the data from the
`/infrastructure/intents.yml` and sends the list of `intents`
into our earlier functions to create the appropriate syntax for
importing our interoperable tokens.

Let's update the `infrastructure/write-files.js` file one last
time to include this process:

infrastructure/write-files.js

```js
import { writeFileSync as write } from 'node:fs';
import { join } from 'node:path';
import { getSchema } from './get-schema.js';
import { getInterop } from './get-interop.js';
import { getInventory } from './get-inventory.js';

import {
  SCHEMA_JSON_PATH,
  TOKENS_SCSS_PATH,
  PUBLIC_DIR,
  INVENTORY_JSON_PATH,
} from './paths.js';
```

This is the complete header for the file. And here is the
introduction of the `getInterop()` function usage:

infrastructure/write-files.js

```
const interopFile = getInterop();

write(
  TOKENS_SCSS_PATH,
  interopFile,
  'utf8'
);
```

There shouldn't be anything surprising here based on our earlier work. Importantly, the resulting file should be accessible to your components such that a developer can import this file in JSX or SCSS to use the token representations within. This means that there could be components alongside this infrastructure, or you could choose to export this as a resource for other teams to use for expressive purposes. Remember, every time the `infrastructure/intents.yml` file is updated, this file will also need to update.

This is the beginning of everything we'll need to start working with this system. In the next chapter, we'll discuss how we'll deliver these resources to the end-user.

Implementation Checklist

The following is a checklist to help verify that all the necessary infrastructure is in place before beginning to author expressions for the system.

Setup

- [ ] Create basic file structure
- [ ] Install necessary dependencies (`js-yaml`)
- [ ] Create the list of intents (`intents.yml`)
- [ ] Set up helpful utilities (i.e., `paths.js`)

Schema Management

- ☐ Generate schema (`get-schema.js`)
- ☐ Generate mode file template (`bootstrap-mode.js`)

Mode Inventory

- ☐ Generate Inventory w/ CSS (`get-inventory.js`)
- ☐ Write `.css` files from modes in public folder
- ☐ Write `_inventory.json` in public folder

Component Resources

- ☐ Generate interoperable token module (`get-interop.js`)
- ☐ Make available to component authors

Endnotes

1. *https://github.com/designsyshouse/mise-en-mode*

2. *https://developer.mozilla.org/en-US/docs/Web/CSS/Reference/Values/system-color*

3. *https://sass-lang.com/*

4. *https://github.com/css-modules/css-modules*

5. *https://sass-lang.com/documentation/interpolation/*

File Delivery

From the previous chapter, we should have all the files we need to start delivering our new expressions to our organization's experiences. In this chapter, we're going to discuss the complexity of managing and delivering the modes.

From an engineering perspective, it would be beneficial to only provide values for modes that are on the current page and separating the modes by purpose would help in this effort. Otherwise, sending all the modes has the possibility of including unused resources on a page. This could slow the network for getting user data, especially with the possibility of dozens of modes. However, separating modes is still a beneficial approach, as this reduces the duplication of font-related intents that would most likely appear similar across several files focusing on color such as `_light.css`, `_dark.css`, and `_promo.css`.

The amount of variables sent to the page may also be a performance concern. However, consider the traditional approach of including tokens for concepts like inverse. Ultimately, we still need to define all of the values to present these expressions. The benefit of Mise en Mode is that we deliver small chunks of values as modes when needed, instead of an entire payload that could include resources a page never needed.

Any approach that conditionally requests a mode should be attempted on the initial request for the application, and optimally no later. While it would be easier from a client-side perspective to listen for new modes to be introduced, styles requested later in the application lifecycle, even with nearly no delay, are subject to a Flash of Unstyled Content (FOUC)[1]. For modes that influence size and spacing, this can cause elements to shift during load and potentially cause a user to make errors in their experience. Instead, it would be beneficial to have styles written with the initial page load. This includes modes and their CSS.

Attempting to analyze what modes are required at the server is also not entirely assured. Reading the served HTML to identify the modes applied will certainly solve most initial load concerns. However, perhaps modes are conditionally added later in the lifecycle of the application, making it more challenging to identify through any process. Especially if the resources are not triggered by the server, but by some client-side activity.

Under these considerations, this suggests that a hybrid approach would be best. The initial load of the page would be analyzed for modes and served with the HTML, and any subsequent modes could be discovered later in the lifecycle by some listeners which are triggered when a node is updated to expect a mode to be applied.

You may be tempted to render these files on-demand instead of collecting all of the necessary files to be served. There's benefits and drawbacks to dynamically creating the resources. Dynamically creating these at an endpoint reduces the number of files needed to be stored. However, it also may increase the time spent fetching the resource, especially if there's any additional processing required. In my opinion, a few more kilobytes of space to make a quicker user experience is the way to go.

Initial Modes

The first part of the lifecycle can happen before a page is ever requested. Unlike the requirements to initially render a page to identify the most important styles, the number of modes we can determine as important should be known. For example, we could be sure that the mode which represents the default brand expression should be on the list. Meanwhile, modes that are less important to appear immediately on page load could be requested later, especially in the case of modes consisting only of color.

Preparing the most important CSS is not anything new. This is often called "critical CSS", and there have been techniques that help identify what that CSS would be. Most of those techniques would require rendering the page, identifying the elements that appear in the viewport at first load, and then matching those elements with selectors found in the CSS. While we *could* identify modes being used in this way, we could also do something more simple. We could agree what the most important expressions are to exist on any page on initial load. Typically, this will be some initial brand expression and perhaps some system expressions depending on how important they are to the overall experience.

If the organization can agree which modes should most likely be available on each page, delivering those modes should be a matter of either writing the CSS into a `<style/>` tag or referencing the external CSS file as `<link/>`. Nothing too tricky here, we can use a bit of JavaScript to write the files easier.

```
<link rel="stylesheet" href="/brand:light.css"/>
```

In this example, I've named the modes and files as `owner:purpose`. We'll discuss more considerations about naming the modes later. This is assuming that the mode CSS files would be found at a publicly accessible directory. If you remember in the earlier chapter, this will depend on where you want to write these files in the `public` folder and how they are eventually served.

The unfortunate part of this is that applications would need to know the file name associated with the mode. It's possible that these don't match due to the needs of the organization or even the restrictions on file naming convention or expected HTML selector.

In the best case scenario, you'll want to keep this list of initial modes small and not load *all* of your modes at once. CSS is a blocking resource. So when the page begins to load, the loading of elements will pause to retrieve the mode CSS before continuing. The more CSS you expect to load, the longer that pause will be. Certainly, determining what the most important expressions are for the business can be a challenging task to complete. There are a few ways of thinking about this.

Each page or application could be responsible for what their most important modes are. This would mean that each team would be supplying some way of choosing which of the available modes should be loaded first. This would mean that all of the modes would need to be available to all applications for that selection to occur. Each team would also need to be aware of the responsibility to provide the initial modes to their application. Not setting the initial modes could result in an unstyled application on initial load. This could be unexpected for poor connections or cause stakeholders to complain that the page doesn't look like it is intended to on first load.

Another way to manage this is for the author of each mode to be responsible for how important it is. This would be an additional configuration key for any particular mode that sets its priority in comparison to other modes. The highest of all of these would be the first of the modes to be loaded. You might imagine that this would be a nightmare to manage. Owners of the modes would then battle to appear on top, in a similar way to how `z-index` numbers in CSS have a tendency to conflict amongst components. Overall, I wouldn't recommend identifying priority in each of the individual files.

What I would recommend, as a start, is for the design systems team to have the responsibility of preloading the initial modes so that any additional modes included could also be written to the page. In this way, other feature teams wouldn't need to be responsible for initial modes unless they wanted to include one. This is similar to the task of a design systems team to be responsible for the agreement among teams as a resource. This is a matter of communicating with stakeholders and understanding the impact of the modes such that the most important ones can be agreed upon and served with each request.

In my opinion, there should be some friction to adding initial modes past what is agreed. It is very possible that if an interface is too easy to use, that the feature is abused without knowing the impact. In this way, I'd suggest that folks looking to introduce modes of importance to perform a documentation lookup to determine how to do this; specifically how to include other modes in the list.

Importantly, I don't recommend that these mode identifiers should be referred to directly with the file names being used, but instead the keyword trigger that would apply the mode in place. As an example in the HTML `data-mode="light"` attribute, the keyword trigger for the mode is `light`. If this aligned to a file name directly, this would assume a `_light.css` file which is possibly not descriptive enough. There could be other files expecting a light motif, especially across brands. Separating the file name construction with the keyword trigger allows for the file naming scheme to be separate from the mode naming. This allows for multiple `light` modes due to having different files for them, among other helpful defining qualities we'll see later.

If you've followed the infrastructure so far, this is the reason why the `mode` key is defined within the metadata of the file, and not determined from the mode file name. We can use this data, along with other logic to determine what mode to serve.

Dynamic Mode Requests

So far, we have enough infrastructure to serve modes to any page. However, it is possible that some components might expect a new presentation later in the application's lifecycle. Imagine our pricing table from earlier as a component that could be on any page. We'll need to also catch when elements expect a mode that isn't considered as important for the initial load. In other words, we need to listen for the introduction of modes that have yet to be covered within the page.

While you might consider using a `MutationObserver()` to check for changes on the client-side, I'd recommend a different approach which is more performant, based on the node insertion detector made popular by David Walsh[2]. The gist is that an animation will trigger on an element that matches a selector. For example, assuming that a `<button/>` element is inserted into the DOM, let's review the following CSS:

```css
@keyframes grow {
    to { scale: 2 }
}

button {
    animation: grow 2s ease forwards;
}
```

This would cause each `<button/>` element to grow as soon as it is inserted into the page. The use of `scale` here is just an example of a CSS animation definition. The trick is to use a script that can listen for CSS animations happening with the following code:

```js
window.addEventListener('animationend', (ev) =>
{});
```

The function will fire each time an element has an animation end on it, for each element in the page. This means we can get a node reference to each element by referencing the `ev.target` just like any other listener.

```
window.addEventListener('animationend', (ev) => {
  // Element that was animated
  console.log(ev.target);
});
```

Inside the function, we can check the element for mode qualifiers. If we're using data attributes to apply a mode (e.g., `data-mode="light"`), we could assume the following to identify which mode needs to be applied on the page.

```
window.addEventListener('animationend', (ev) => {
  const { mode } = ev.target.dataset; // "light"
  // Request the resources for this mode
});
```

> ⓘ **NOTE**
>
> You'll want to check that a mode was actually supplied by the node. Remember this function will trigger on *all* animations.
>
> Be careful when qualifying with `ev.animationName` because it's possible the element with the `data-mode` could have a visual animation playing, which could override the detection CSS.

Now we need some CSS that is specific to modes. Using the CSS data attribute selector, we can target all elements that have `data-mode` assigned and cause them to run an animation:

```
@keyframes _nodeinserted_ {
  to { visibility: visible }
}

:where([data-mode]) {
  animation: _nodeinserted_ .0001s linear forwards;
}
```

Using the `visibility: visible` declaration makes the animation trigger by using a rarely-used property that shouldn't conflict with other more common properties. This works even with the default value matching the new one provided by the animation. The use of `:where()` reduces the specificity to 0, so more important animations can still play but also trigger the script.

Using this logic, we can detect when an element introduces a new mode and the page is completely loaded and which mode it is so its resources can be requested as needed.

Lazy Mode Requests

At this point, we've covered how we might preload modes that are most important to be included on the page, and modes that could appear during the lifecycle after initial load. One more case that might be helpful is to get any remaining modes if resources are available. This might be important because the data could be cached on the client-side in cases where resources might not be available later. When the user navigates to other pages, we can be sure that the mode resources are available even if other resources might not be.

What makes this complex is determining the threshold for allowing additional resources. As an example, you might want to only get remaining modes if the network connection is fast. However, we also don't want to be too eager to request a mode if the user is on a cellular data plan. How you optimize this is up to you, but having empathy here for all users and their connection types is important to keep in mind.

One way we might consider doing this is by waiting for the browser network to become idle and then checking for a quality connection. At the time of this writing, the following scripts aren't fully supported in browsers, but this could give you some ideas on how you might want to introduce the last modes not yet found on the page.

```
if (typeof window?.requestIdleCallback ===
'function'
  && navigator?.connection?.downlink >= 10)
  requestIdleCallback(() => /* Request remaining
resources */);
```

Another way is to update the `public/_inventory.json` metadata to include the size of the CSS file for the mode, and then time how long it takes a file of that size to be returned. Adding the file size to the inventory can be done in the loop within `getInventory`.

```
return acc.concat({
  mode,
  css,
  href: `_${name}.css`,
  bytes: Buffer.byteLength(css, 'utf8'),
  ...metadata
});
```

How you determine if it is appropriate to request the remaining modes to the client will be up to your organization, including if this enhancement is necessary at all. For organizations that have a small amount of modes to maintain, this might be an over-optimization.

Mode Manager

Let's take all of the concepts we've gone over and create a system to support the lifecycles a mode can be requested within. The expectation is that all of the code below is meant to go into a single `mode-manager.js` file.

mode-manager.js

```javascript
function customToString(target, value) {
  Object.defineProperty(target, 'toString', {
    value,
    enumerable: false
  });
  return target;
}
```

This is a helpful function which overloads the given object's `toString` function. We saw something similar within the `getInventory` function earlier, when we needed to omit the CSS from the `inventory` after writing as JSON. This function allows objects to be passed around with their metadata intact, while also writing the representation as a custom string, like writing an HTML element from a collection of object properties.

mode-manager.js

```javascript
function headAppend(tagName, attrs) {
  const $elem = document.createElement(tagName);
  Object.assign($elem, attrs);
  document.head.appendChild($elem);
}
```

This function will only be used on the client, to create and append the given `tagName` and `attrs` for that element to the `<head/>` of the page. This is done because `<script/>` tags do not execute unless they are appended as nodes. In other words, writing a `<script/>` using `insertAdjacentHTML` would cause the contents to be inert and never execute.

mode-manager.js

```javascript
function linkify() {

  const inventoryPromise =
import('./_inventory.json', { with: { type: 'json'
} }).then(mod => mod.default);

  return async (mode) => {

    const inventory = await inventoryPromise;

    const { href } = inventory.find((entry) =>
entry.mode === mode) || {};

    return customToString({
      href,
      rel: 'stylesheet',
      title: mode
    }, function () {
      const toAttrs = ([attr, val]) =>
`${attr}="${val}"`;
      const attrs =
Object.entries(this).map(toAttrs);
      return `<link ${attrs.join(' ')}/>`;
    });
  };
}
```

OK, this function might look a bit gnarly. The `linkify` function produces a closure (later found in this file as `toLink`) that is used to generate `<link/>` tags from a given `mode`. We'll see this in action in later functions.

The reason why it is set up this way is to solve for a race condition. The `_inventory.json` import could resolve later than the first node insertion, causing the `inventory` to not be available for the resource lookup. This would cause the first modes found to not be added to the page until possible later additions. When this function is first used, it is preloading the request for the `inventory` before it is needed. If it is needed quickly, we'll `await` for the resulting function before finishing the node insertion callback.

The rest of this function should make sense. We'll find the `href` from the given `mode` in the inventory. If it exists, we'll return the necessary attributes to create a `<link/>` tag, along with the overloaded `toString` function that will automatically apply the given attributes to the element when stringified.

mode-manager.js

```js
function observer(preload = '') {

  const loaded = new Set(preload.split(','));

  const toLink = linkify();

  const range = document.createRange();
  range.setStartAfter(document.currentScript);
  range.collapse(true);

  window.addEventListener('animationend', async
(ev) => {

    const modes = new
Set(ev.target.dataset?.mode?.split(' ') || []);

    const diff = [...modes.difference(loaded)];

    diff.forEach((mode) => loaded.add(mode));

    const links = await
Promise.all(diff.map(toLink));

    links.forEach((link) =>
range.insertNode(range.createContextualFragment(lin
k)));
  });
}
```

Another gnarly few lines of code. The first line expects the input to be a comma-separated list of modes that were preloaded. The reason why this isn't given as an `Array` will be clear later. We make this list into a JavaScript `Set` which has an important feature: only unique entries are kept within. This ensures that we don't store duplicate entries, meaning we shouldn't load the same mode twice. We also prepare the `toLink` function responsible for making `<link/>` elements.

The next block is a unique way of inserting elements into the page. Effectively, we're creating a pointer that tells later logic to place elements at this specific location. In our case, the location is directly after where this current script is placed. While you could omit this complexity, I find it helpful to keep resources related to the mode together in the `<head/>`. Otherwise, resources loaded later might also be found later in the `<head/>`, making it hard to catch what is being added over the lifecycle of the page.

The listener added here should be familiar from the earlier explanation. We're going to identify all of the modes placed at this element. This is a bit more robust than our earlier examples, capturing the possibility of multiple modes on a single element. Next, we use the `Set.difference()` method which helps us determine which modes identified have not yet been loaded. We'll add those modes to the list and also create `<link/>` elements for them. Finally, we'll use the pointer from earlier as the place to insert the given `<link/>` resources which will request the new mode CSS for the page.

mode-manager.js

```js
function styleAttrs() {
  const animationName = '_nodeinserted_';

  const textContent = `
    :where([data-mode]) {
      visibility: hidden;
      animation: ${animationName} .0001s linear
forwards;
    }

    @keyframes ${animationName} {
      to { visibility: visible }
    }
  `;

  return customToString({
    textContent
  }, function () {
    return `<style
type="text/css">${this.textContent}</style>`
  });
}
```

This should be mostly familiar from our earlier understanding, with the additional need to use the `customToString` function to create the `<style/>` tag. Remember, this is used to support isomorphic rendering. The returned object will immediately create the `<style/>` tag if stringified.

mode-manager.js

```javascript
function scriptAttrs(modes) {
  const args = modes.map((mode) =>
`'${mode}'`).join(',');

  const textContent = `
    ${customToString.toString()};
    ${linkify.toString()};
    (${observer.toString()})(${args});
  `;

  return customToString({
    textContent
  }, function () {
    return `<script
type="text/javascript">${this.textContent}
</script>`;
  });
}
```

This function stringifies some of the earlier functions written in
this file so that they can be used on the client. This is the reason
why the `observer` function accepts a string as the list of modes.
Because it is easier to `split` and `join` on an array of strings
instead of `Array`-ifying the `[]` syntax from a string. The
`customToString` function is used within `linkify`, making it
needed for the client for creating the `<link/>` HTML strings.
Finally the `observer` is rendered within an IIFE[4], causing it to
execute immediately once it is written.

Writing the function this way, as the `toString` representations
of actual JavaScript functions, helps with maintenance of this
source code. Otherwise, writing these as strings would miss out
on enhanced developer experience such as syntax highlighting
in a code editor.

mode-manager.js

```javascript
export async function modeManager(preload = []) {
  const modes = [].concat(preload);

  const toLink = linkify();

  const html = {
    link: await Promise.all(modes.map(toLink)),
    style: styleAttrs(),
    script: scriptAttrs(modes)
  };

  if (typeof document?.head !== 'undefined') {
    Object.entries(html).forEach(([tagName, shape])
=> {
      [].concat(shape).forEach((attrs) =>
headAppend(tagName, attrs));
    });
  }

  return Object.values(html).join('\n');
}
```

This is the `modeManager` function export which is responsible for creating the initial resources to be set on the page isomorphically. The first line allows a `string` input or an `Array` of mode aliases meant to be present initially before later lifecycles. We also prepare the `toLink` function meant for the initial modes.

The next object represents the HTML to be written onto the page. Remember, these functions create isomorphic objects from the `customToString` function. So, when these values are stringified, it creates the HTML string meant to be added to the `<head/>` of the page. This is precisely what is returned in the `modeManager` function in the last line.

Before that, we can check if this is being run in a browser context which would include a `<head/>` element. If it is, then this was meant to take the given HTML and have it appended to the `<head/>` using the `headAppend` function from earlier. The reason why there are two `forEach` loops is to handle the possible

collection of `<link/>` elements. We first loop through the initial
values of the `html` object, and then loop through the `Array` result
found at the `link` key in a similar way.

Lastly, we have a final line in the file:

mode-manager.js

```
if (typeof window !== 'undefined') {
  modeManager();
}
```

If you include this in the browser, this will instantiate the
manager immediately. This is also where you can provide the
initial modes, either as a `string` or an `Array` of mode aliases.

```
modeManager('light');
```

Due to the use of dynamic importing, when including this script
client-side, the script must be introduced as a `type="module"`.
Here's what that might look like:

```
<script type="module" src="mode-manager.js">
</script>
```

> ⚠️ **WARNING**
>
> If you are working in an environment that uses the Shadow
> DOM, some of the logic in the manager will need to be
> updated. Namely, checking and appending to the `<head/>`
> (since the Shadow DOM does not have a `<head/>`) and
> where to attach the listener for client-side modes (since the
> Shadow DOM does not share the `window` scope, it is
> blocked).

Where you place the `mode-manager.js` file is dependent on your
application setup, specifically whether you can render resources
on the server or client. If you expect this to be used only on the

client-side, then you would place it in a location like `public` from our earlier infrastructure. In a server-side architecture, you'd want to import the `modeManager` function in a place that renders other resources for the `<head/>`, hopefully in a reusable way.

Chapter Summary

In this chapter, we explored the challenges and strategies of delivering modes efficiently. We established that while separating modes by purpose reduces duplication and improves performance, care must be taken to ensure the right modes are loaded at the right time. Initial modes should be prioritized to prevent visual glitches like FOUC, while additional modes can be requested dynamically or lazily depending on the needs of the organization.

We also saw how these ideas come together in the `mode-manager.js` file, which provides an isomorphic way to preload essential modes, detect new ones during runtime, and manage the full lifecycle of delivery. The key takeaway is that a balanced, hybrid approach—combining initial server rendering with client-side detection and optional lazy loading—creates the best experience for users while maintaining organizational flexibility.

With our delivery system in place, we can now focus on how to effectively curate and manage the values within our mode files.

Initial Modes

- Load only the most important modes during the first render to prevent FOUC and layout shifts.
- The design system team should manage initial mode priorities to ensure consistency across applications.
- Keep mode keywords (e.g., `light`) distinct from CSS file names for flexibility and scalability.

Dynamic Modes

- Combine server-side detection of initial modes with client-side listeners.
- Use animation-based detection for inserted nodes.
- Fetch additional modes only when conditions allow (e.g., idle browser, strong connection).

Mode Management

- The `mode-manager.js` has the sole responsibility to preload, detect, and manage modes throughout the lifecycle of a page.
- It is provided isomorphically, where it can be used on the server or within the browser.
- Additional enhancements may need to be considered for strict environments, such as using Shadow DOM.

Endnotes

1. *https://en.wikipedia.org/wiki/Flash_of_unstyled_content*
2. *https://davidwalsh.name/detect-node-insertion*
3. *https://en.wikipedia.org/wiki/Isomorphic_JavaScript*
4. *https://developer.mozilla.org/en-US/docs/Glossary/IIFE*

File Authoring

By this point, we should have a significant amount of infrastructure in place to support design tokens and modes. Now that we have this, it's time to finally start using what we've made to express creativity in our experiences.

Curating

A big part of Mise en Mode is choosing the values meant for each intent to create a single mode; the exercise of **curation**. Teams can have varying degrees of tooling to help with choosing these values. More sophisticated teams may have an entire internal user interface dedicated to the process of choosing tokens that makes identifying changes clear in a visual sandbox environment. A color picker would allow a designer to choose a value quickly for any token and watch how it propagates through the dynamic wiring of styles on a page. Teams with fewer resources may be confined to editing a text file by hand, where seeing what is affected by any token value change is a cumbersome process.

No matter where your team might lie within this range of possibilities, I believe you should prepare for the worst-case scenario. Assume there is no tooling and that files are manipulated by hand. This means that they should be in a human-readable state where some later processes could easily ingest the data as styles later down the line. This is especially helpful in the case where a new token or concept is introduced and waiting for an update to the editing experience is a waste of valuable time. Furthermore, there is also a benefit to being able to comprehend a file of data without the help of some additional tooling. Think about any time you needed to download a program just to open a file, while other files can open as intelligible plain text. This is a large reason why `.yml` was introduced earlier as a replacement for other data structure syntax options.

So, if we are thinking about the worst case and there is no easy way to verify how a change affects the experience, how can we gain the confidence for a well-crafted mode we need during curation?

First, consider the file that represents a mode. Perhaps this is "light mode", and inside exists a configuration of intents and values that are meant to represent what the user would see when these values are applied within a light mode. This means that we aren't thinking about dark mode—or any other mode— for that matter. The values for light mode and dark mode should never coexist through tokens; *all values exist within the boundary of a single mode.* This is important because all of the decisions within a single mode are meant to convey that one expression. Imagine the difference in playing 12 opponents in chess, where in one event you can start and complete each game before moving on to the next opponent. Now consider an event where you'd switch opponents after every move. That context-switching makes it harder to manage all of the variables in any game. It's considerably easier to finish one game before moving to the next. The same concept is applied here when curating. It is easier to complete one mode at a time instead of trying to span value decisions across multiple modes.

This may be in contrast to other token structures that you may have seen. As an example, a popular approach is to name the token, and then list all of the value variations underneath as a collection. Here's what that would look like in JSON notation:

```json
{
  "$surface_auxiliary_backgroundColor": {
    "light": "white",
    "dark": "black"
  }
}
```

This is also why I feel some of the nested object structures in example token files are more noisy than helpful. Having values representing light in the same file as dark overwhelms the curator in the value curation process. While it might not matter if computers are reading this file, I still believe that there's a benefit to a human-readable format.

> ### (i) NOTE
>
> For the following examples, I'll be illustrating the *output* of the architecture as CSS, instead of the expected YAML syntax. I've chosen to do this so you may check for the expected results of the infrastructure supporting the curation of these values for accuracy. It's also a more familiar syntax to folks styling web pages than the newly introduced YAML.

Using the one mode per file approach, within that one file, there is a single selector that identifies when the collection of values should be applied inside of a context. Based on the architecture we've completed so far, a `_light.css` file might look like this inside:

```css
[data-mode~="light"] {
  /* CSS custom properties */
}
```

Within that selector, there is a list of tokens. But not just any tokens; strictly intents. In our example, this would be a list of

CSS Custom Properties and their chosen values in the syntax
we used earlier that includes the emoji:

```
[data-mode~="light"] {
  --🔒surface_auxiliary_backgroundColor: white;
  --🔒surface_auxiliary_foregroundColor: black;
}
```

In a completely separate file called `_dark.css`, we would have an
identical list of intents under a different selector with different
values assigned to describe the dark mode as different from
light mode.

```
[data-mode~="dark"] {
  --🔒surface_auxiliary_backgroundColor: black;
  --🔒surface_auxiliary_foregroundColor: white;
}
```

The expectation between these files is that they must have
precisely the same intents across all mode files. The values can
—and probably should—be different, but the intents are the
same. Remember that intents suggest how a stakeholder means
to express themselves in a functional experience. The function
of an experience should not change when we apply a new
expression, therefore the set of intents we use should also
not change.

Now, a designer can choose where `data-mode="dark"` should be
applied within the page. This could affect the entire page by
placing it on the `<body/>` or, in the case described by Mise en
Mode, a part of the page like `<footer/>`.

```
<body data-mode="light">
  <footer data-mode="dark"></footer>
</body>
```

Assuming that the `_light.css` and `_dark.css` are both loaded
on the page and the tokens are appropriately assigned to the
style properties within these areas, we should see the `data-`

`mode="light"` area receive a white background with black text, and the `data-mode="dark"` area receive a black background with white text.

Earlier, we suggested that a mode is an expressive enhancement to a visually simple blueprint of a user experience: the decisions we make about how any part of this structure could be determined through a collection of intents assigned to those elements that describe the parts, where the values come from a single mode that describes the expression. In these initial explorations, we've limited the values that are assigned to a single mode: the background for `_light.css` is white, while the background for `_dark.css` is black.

Separating Mode Concerns

In our examples thus far, we've been primarily using color to demonstrate the mode change. This is because many of us are familiar with the expectations of light and dark modes and using color is the most striking visual change. Based on what we've discussed in previous chapters, we aren't limited to only color, but any token that can appropriately exist as an intent—such as text.

Let's think about how we might support text with what exists so far. We've said earlier that a mode should have all the intents listed within each file so that full coverage is guaranteed, no matter what mode is applied. This would mean that the list of intents would not be restricted to only color but all intents that would appropriately be defined. This means we might have the following content, abridged for clarity:

```css
/* _light.css */
[data-mode~="light"] {
   -- 🔒 surface_auxiliary_backgroundColor: white;
   -- 🔒 surface_auxiliary_foregroundColor: black;
   -- 🔒 text_auxiliary_fontFamily: Roboto;
}

/* _dark.css */
[data-mode~="dark"] {
   -- 🔒 surface_auxiliary_backgroundColor: black;
   -- 🔒 surface_auxiliary_foregroundColor: white;
   -- 🔒 text_auxiliary_fontFamily: Roboto;
}
```

We'll see that the text intents and their values will be duplicated
across these two modes. You might argue that this is redundant,
and I would agree. It would be more organized to separate the
text intents from the color, grouping closely related properties. It
makes sense to use the same font for light, dark, and
subsequent different expressive enhancements using color. Our
expectation is for colors to be updated more frequently than for
fonts to be redesigned—especially from a branding and
marketing standpoint. In this way, it makes more sense to have
separate files for color, typography, and other stylistic properties
where we expect to update the expressions:

```css
/* _typography-roboto.css */
[data-mode~="roboto"] {
   -- 🔒 text_auxiliary_fontFamily: Roboto;
}

/* _color-light.css */
[data-mode~="light"] {
   -- 🔒 surface_auxiliary_backgroundColor: white;
   -- 🔒 surface_auxiliary_foregroundColor: black;
}

/* _color-dark.css */
[data-mode~="dark"] {
   -- 🔒 surface_auxiliary_backgroundColor: black;
   -- 🔒 surface_auxiliary_foregroundColor: white;
}
```

What this suggests is that it might be more appropriate to separate modes by purpose, similar to how we separate tokens by purpose as intents. This means that `_color-light.css` might only include color changes and a separate file `_typography-roboto.css` describes changes that affect the typeface. This way, we can reduce the number of values we'll need to curate for expressive enhancement. As an example, we can introduce `_color-promo.css` to describe an area that enhances the messaging for a promotional deal with new fun colors without needing to include sensible defaults for the font properties. The typography values can be served separately even while the experience expects a wholly complete expression. For example, if you believe the font weight should change between light and dark modes, that would constitute a `_typography-light.css` and a `_typography-dark.css` served with the other appropriate modes for full coverage.

This also demonstrates the benefit of separating the file name from the keyword trigger. The file names could be constructed in such a way for easy organization, while the mode keyword trigger can be minimal for use within the page. This allows for multiple keyword triggers to appear at a single location in the experience, each of them handling a different expression concept.

Another benefit to this separation is that folks curating aren't responsible for all the values where providing them would be redundant (or worse: incorrect). Realistically, the changes we expect to happen most often are color-related. Changes in font or other properties are less likely to need change over the lifecycle of a product platform. Separating the files separates these concerns. The `_typography-roboto.css` could have stricter editing permissions than modes that primarily handle color.

Verifying Coverage

Using this separated approach is somewhat risky: you must ensure that values for all intents are applied to every page, but

not every mode would guarantee all intents are provided. As
mentioned earlier, if this mode is a human-readable file, it is
more likely for a value to be missing or omitted entirely by
human error, causing a value to not propagate through the mode
at all. Ultimately, this can easily cause a visual regression. To
establish quality using this method, you'll want some validation
steps to ensure all the expected intents exist based on the mode
separation. Sensibly, validating coverage would be helpful
regardless of the approach, but it is more critical when
separating modes by purpose.

Here's how you might begin to do this when constructing your
inventory file with `get-inventory.js`:

infrastructure/get-inventory.js

```javascript
import { readFileSync as read } from 'node:fs';
import { load } from 'js-yaml';

import { INTENTS_YAML_PATH } from './paths.js';

const intents = new
Set(load(read(INTENTS_YAML_PATH, 'utf8')));

function getCoverage(tokens) {
  const check = new Set(tokens);
  return (intents.intersection(check).size /
intents.size) * 100;
}
```

These additions are loading the `intents.yml` once so that the
loop doesn't continuously load the YAML file. The file's list of
used tokens is made into a `Set` so that we can use the
`Set.intersection()` method to get the percentage of intents
the given file uses. Here's how we'd update the loop to include
this information:

infrastructure/get-inventory.js

```js
return acc.concat({
  mode,
  css,
  href: `_${name}.css`,
  coverage: getCoverage(tokens),
  ...metadata
});
```

Then the `modeManager` can be updated to warn if an initial mode isn't at 100% coverage. When creating the `toLink` function from `linkify`, add the `coverage` data as an additional data-attribute. Then loop through the returned `<link/>` object representations and check that each has 100% coverage.

mode-manager.js

```js
if (html.link.some(({ coverage }) => coverage <
100)) {
  console.warn('Initial modes may not be fully
covered');
}
```

You might see a problem with this when considering separated modes. So far, this only provides a percentage against the full list of intents for each mode. It is possible that the initial modes *combined* would produce 100% coverage. The issue is that we do not know which modes are initial until the `modeManager` is used; well after the `_inventory.json` is created. This means that the `getCoverage` function will need to be more complex and the data it returns would need to be larger so that the comparison is completed within the `modeManager`.

How you want to calculate coverage comparisons will be largely dependent on how you expect to separate modes. Using our previous example, one of the ways you could do this is by categorizing by property within `getCoverage`. This means that you'd calculate coverage by determining the completeness of `color` in a mode, separate from the coverage of `typography`. The result of `getCoverage` would then be an object showing how much of each category is covered. Here's how that might look as `_inventory.json`:

public/_inventory.json

```json
[{
  "mode": "light",
  "href": "_light.css",
  "coverage": {
    "color": 100,
    "typography": 0
  }
}]
```

Assuming this object is included within the `toLink` function result, then in the `modeManager`, we do a more specific comparison against all of the incoming `<link/>` element objects, looking for 100% coverage.

mode-manager.js

```js
if (!html.link.some(({ coverage }) =>
coverage.color === 100)) {
  console.warn('Initial modes are not fully covered
for color');
}
```

If we separate the modes based on style and expected frequency of editing, having a system like this in place will be beneficial to check that an experience is set up for a successful expression on first load. Remember, checking coverage is an optimization to help identify possible gaps. If you don't separate the contents of modes, you could avoid checking for coverage entirely.

Compounding Influence

Assuming that we're well-prepared to support an optimized separated approach, let's return to focusing on the act of selecting a color. Specifically, what goes through our mind when choosing a color for an intent in the realm of visual design?

The first thing that might come to mind is the brand. This is reasonable as we certainly want whatever color we choose to reflect the perception of the company and drive trust that this value is part of a larger whole. From here, perhaps we're supporting light and dark mode, so we'll want to adjust whatever color we've selected to align to that mode. This might not always have such a large impact. We'll focus on using brand colors for accents and less interesting colors to support surfaces and their content.

What about that promotional mode we considered in an earlier example? Certainly, there are brand colors involved, but there would also be the user preference for light or dark mode. So we'll need to make some decisions using three things in this case: brand, user preference, and promotional enhancement. Recognize that juggling these influences is happening for each token you curate because these modes do not exist in isolation

—they are compounding. Each value being decided is meant to be influenced by a few or more factors which would make `brand-light-promo.css` the more appropriate file name. Let's consider what this might look like as marked within HTML:

```
<body data-mode="light">
  <section data-mode="promo">
    <!-- card -->
    <!-- card -->
    <!-- card -->
  </section>
</body>
```

Light page with a promotional section that isn't light

In this example, `promo` doesn't indicate anything about it being a part of an overall "light" or "dark" page. The whole context of that `<section/>` is contained and may not have any influence outside of it. In fact, this is a *benefit* of the approach: these areas don't conflict with each other. Does this mean we'll need to define `promo-light` and `promo-dark` to have more consistency between modes? We'll revisit this example for a possible answer later.

Even with three factors, the number of possible files and mode combinations could explode quickly, requiring a decision to be made for every token inside. While we're thankful this is reduced to color, there remains a lot to curate.

Further complicating the approach, brand, user preference, and promotional enhancement isn't an exhaustive list of influential factors. Culture may play a role in what color to present. As an example: in China, the color black is not appropriate for a button's background since the color is associated with mourning and misfortune. While you certainly could consider changing your overall brand approach to avoid this particular cultural faux pas, the change may not account for others. This means our color decision may also include a mode influence of culture. This would introduce further mode permutations such as `_brand-light-chinese-promo.css`. This also doesn't yet cover the possibility of showing a critical or selected state within this mode.

This puts a great deal of pressure and responsibility on a person trying to account for all the factors when selecting a single value or creating an entirely new expression. Making it pop could become an exploration of socioeconomic design that could seem wildly tangled. It doesn't seem reasonable for a person to be responsible over the best value for the eventual compounding expression they are about to create. Many unknowns will exist.

This is often where we stop. It's too complex for any one person or even a single team to consider all these permutations. We'll pick a value that is good enough until it's not, and maybe we'll reassess it in the future if there's enough squeaking on this particular wheel. The `_color-light.css` and `_color-dark.css` files will totally use our one collection of branded colors. We'll make some special tokens for promo components and hope we don't offend anyone with the colors we chose. It's true, this traditional approach has worked for many teams under our current community's understanding and resources.

But what if we could deliver a better experience without needing to focus on all of these influential factors? That's what the final few chapters will explore. How we might rethink the curation of these values while still enhancing our experience in creative ways.

Chapter Summary

As we've seen, managing compound influences in mode systems presents significant challenges. Specifically, how curating values can be overwhelming for a number of reasons. The next chapters will explore innovative approaches to handling these complexities while maintaining system manageability.

Key Concepts

- Mode files should be human-readable and maintainable
- Separating modes by property type reduces complexity
- Coverage verification is crucial for separated modes
- Compound influences create exponential complexity

Best Practices

- Keep mode files focused and purpose-driven
- Consider implementing coverage checking
- Consider all influence factors
- Plan for mode composition

Future

Inclusive Systems

In the previous chapter, we discussed the combinatoric problem of curating values for intents and modes when many expressions are expected. We identified several factors that could influence the final value for any intent and mode combination. Now let's try to make this exercise of expressively enhancing an experience as accessible as we can by meeting our users' unique expectations.

Inclusive Design

In 1994, design activist Roger Coleman published "The Case for Inclusive Design," which shifted the focus away from universal design to **inclusive design**. The difference is important: the former attempts to create a solution for as many people as possible, while the latter identifies folks who would be outside of the realm of the former by customizing experiences to the needs of those subjects. In other words, inclusive design is attempting to support *everyone* by addressing each individual's personal needs.

You might imagine that aligning to a universal design paradigm is easier to achieve than one that changes for an individual. However, remember the earlier statistic regarding translated experiences. The more folks who can use your product, the more revenue you can collect. So investing to reach more people has the potential for growing a business further than could be possible otherwise. Remember that even people who are unable to use a mouse or see a screen could still be a paying customer. Alienating people with money from purchasing your product is a bad business practice.

This means that it would be in our best interest to provide expressions that are personalized to the user. This could include accessibility considerations such as color contrast or legible font choices. This could also include cultural facets such as using appropriate colors or changing the density of information. There could even be further factors specific to individual users that we have yet to provide settings for. Have you ever wished that you could change something in an interface to make it more usable for you? What if it was not only possible, but happened automatically?

Of course, there is interest and the ability to provide some sort of customization, as we have settings that support accessibility built into operating systems. There are ways to translate content on the fly or provide alternative methods of input. Some platforms, like the Arc browser by The Browser Company, went so far as to provide aesthetic controls to nearly any website in the form of Boosts[1]. These are user-driven changes made to websites that are outside the control of the page. Users could update the colors of their search engine to match their favorite sports team or change the font to something more fantastical. This puts the power of customized expression directly into the hands of the users engaging in these experiences.

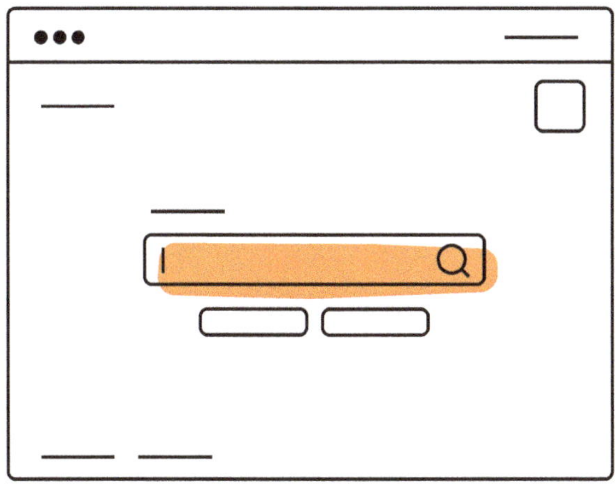

Your most used webpages become more meant for you

Some designers learn this and grab their pitchforks, ready to take back their craft and force "well-designed" experiences back onto their users. These are folks who believe good design only comes from good designers. They hold the opinion that it takes folks passionate about design years to learn what the properties of the best buttons are. On the contrary, the best person to design an experience is the person who is meant to use it. The gap that exists is commonly a lack of resources for a user to execute that perfect experience. It takes a long time to learn what makes good universal design, but it only takes a moment to identify what the flawless inclusive design could be for yourself.

Of course, your perfect experience is most likely not wholly reusable for another person. Nor could we expect that all people have the ability or desire to do the work required to customize each of their experiences. We only expect the world to work for us as it has before. Considering changes for every single person who exists or will exist to have the ultimate inclusive coverage seems like an impossible task. It was hard enough to determine what color the primary button is when only considering the brand and light or dark mode. Trying to include additional personalization factors when curating each value for the final button composition seems too much to consider. Meanwhile, giving direct control seems too flexible or perhaps a distraction from the main purpose of the product. Perhaps we might not provide the best controls for customizing, and a user may leave before ever using the primary feature.

Therefore, the mission is to provide inclusive design decisions to appropriately enhance our experiences without being overwhelmed by the number of decisions any person might need to make. In other words, how can we personalize an experience without requiring that the user take extra steps toward that personalization?

Personalized Values

What we are beginning to discuss is less about finding specific values for any individual expression and more about learning personalization behaviors, using them to influence the experience. As an example, the Nest Learning Thermostat[2] was advertised to gather data based on household behavior to eventually make decisions about how to control a home's climate in the future. This means that the device was personalized without actively configuring the system's settings. It learned through understanding user behavior until the user no longer needed to educate the system. The climate is maintained and personalized to the household.

To fully conceptualize the idea, let's again consider light and dark modes. Traditionally this is an operating system setting controlled by the user inside their device's operating system that may affect the presentation of a website or application if configured. It is also a common practice for these individual experiences to offer custom controls that update the presentation, potentially overriding the setting found in the operating system. The control could include forced light mode, forced dark mode, and observing the system setting. The purpose for exposing this customization control is because the content may be more appropriate in a different mode from what the user generally uses. For example, a user might prefer to use dark mode for most experiences but may feel that reading is easier in light mode. Therefore, the user would prefer all long-form articles to be presented in light mode. Advances in personalization have provided further customization by introducing a feature called "night shift" which adjusts the interface at dusk to shift colors to a warmer side of the spectrum aimed at making the experience easier on the eyes.

We can see that this learning would be most effective at the operating system level, or perhaps some other universal identifier that maintains information across the user's devices, learning behaviors with every interaction. This information would need to be standardized into data that websites and applications could use to further personalize their experiences. As an example, the data could identify what visual treatment this user responds most favorably towards for marketing purposes. A website could then alter its presentation to align with the qualities of this treatment and hopefully increase sales. Remember while we're interested in inclusive experiences, the goal remains to grow revenue by providing solutions that meet user needs and occur through personalization.

What we've described so far is something like `prefers-color-scheme` on steroids. It's reasonable to assume that some system that represents a user would be the mechanism that gathers the data on behaviors, either by explicit user settings or implicit learning of user behavior. Importantly, the operating system would only partially influence the final results as we expect the individual experience to have other expressive enhancements to be considered. In other words, it is most likely inappropriate for every Android application to look like Material Design as it would be difficult to disambiguate which application you are currently in compared to any other. The unique presentation of a brand is important to separate it from the competition, so we want to balance the personalization data provided by the user's system with the identity data of the brand's system.

Fingerprinting

When speaking about hyper-personalized experiences, we'll also need to discuss the topic of privacy. **Fingerprinting** is the ability for a system to identify a specific user based on their settings. You may be familiar with some of the data points which are commonly found in analytics tools. We will often see insight into our visitor's device size, browser, geolocation, among other identifiers. Fingerprinting collects further data to specifically identify an individual in order to track their behavior across unrelated experiences. With many data points, we can uniquely identify a user. This is typically used for targeted advertising, but it could also be used for malicious purposes.

This means there's a fine line we need to consider when introducing the possibility of hyper-personalized experiences. The challenge we'll want to address when considering how we potentially expose user settings is how to keep those settings private.

For example, if we allowed users to provide their own button color for websites, then we could get the computed value for that color and store it as a data point. Then if we see that value on a button elsewhere, we could have some certainty it might be the same user. With more data points, we could further pinpoint the specific user due to the way certain user interface compositions look.

Ultimately, the concept of fingerprinting is in direct conflict with creating inclusive experiences, since any particular experience would need to know the user's settings in order to provide a customized experience. That same data could be used to identify the user—especially since each user's needs are often specific to the user. Some systems attempt to prevent fingerprinting by returning standardized values when scripts are meant to read these settings. In other words, when getting a computed value for a color, the result is always the same. Even if the user has updated this value to be something specific and personal. This doesn't mean that the value isn't displayed how the user expects, it means the underlying value cannot be accessed directly through normal scripting.

In the coming examples, we're going to introduce the concept of System Color[3] as a way to read the color of a user's settings in order to help alleviate the need for deeper curation. In practice, the values provided by these system colors are currently in various degrees of support and stability. However, the theory behind them is interesting enough to warrant the possibility of using them in an exploratory fashion.

System Influence

When the user chooses light mode, this indicates that the resulting color is meant to have most of the responsibility placed on the system's setting, *not the brand*. Certainly, before this approach, we would have to keep the concept of "light" in mind while curating the color along with any other possible modes that would influence the final value. However, note that while we hold the concept of light during curation, that is ultimately guiding our decision to choose the color. Without the concept of light, we could choose any color within the brand's palette without restriction.

In another example, let's consider how this might be used to convey promotions instead. Because the concept of a "promo" is an expression provided by the owners of the application, we expect less influence from the operating system which doesn't need to describe the concept of promotion. An operating system is not meant to sell you products. Brands sell products. We want the values of properties like color to be determined more prominently by the brand. It doesn't mean that the system value is totally absent, it means it has less influence on the final value.

In fact, this can be simplified to two kinds of colors being decided. There's what the system is providing and how the brand wants to adjust that setting. This seems counterintuitive from our previous examples that introduced mode files such as `_brand-light-chinese-promo.css`. It would seem there's several kinds of influence that could happen. However, we can group these into two buckets: values that are determined by what the user wants, and values that are determined by what the brand wants. Reviewing that earlier file name, the concept of expecting a lightened color presentation with Chinese cultural support are both settings that we'd expect to come from the user's operating system. Meanwhile, the brand wants to convey its own concept of treating this part of the interface as promotional.

In a more colorful example, the brand is expecting to use `purple` as the button's background. However, they are also expecting this final value to have a large influence from the system's color for this element. This would cause whatever color is used in the operating system to be shown closer to the `purple` provided by the brand.

You should feel uneasy about this, because the operating system would also introduce its own hues—a mix of an uncontrolled hue could result in an awful presentation. If the user's settings were to provide an `orange` hue for their buttons, but expect that to be mixed with the brand's `purple`, we'd most likely see something that is effectively unbranded, having either too much of a mix or none at all.

What we really want is to use some parts of the system color to influence the brand color. In other words, if the system is expecting to present a button in light mode, we should be replacing the hue provided by the system with the brand hue. We want to keep the lightness and contrast of the system. This would require us to separate the parts of the color.

We can actually explore this as a possibility today:

```
[data-mode~="promo"] {
  -- 🔒 action_primary_backgroundColor: color-mix(
    in oklch,
    oklch(from purple none c h),
    oklch(from ButtonFace l c none) 80%
  );
}
```

This introduces the CSS color-mix()[4] function that allows colors
to be mixed together, and CSS relative color syntax[5] which
allows you to split the parts of the given color into channels. This
allows us to take the lightness (the intensity of light) and chroma
(the difference from gray) from the system color, and mix with
the hue (the color) provided by the brand by some amount of
influence. The higher percentage of influence, the more the
system is in control of the final value. In the above example, the
brand is saying that 80% of the color should come from the
system's color, provided by whatever is set as `ButtonFace` by the
operating system. This makes 20% of the resulting color to be
the `purple` provided by the brand.

> ### ⓘ NOTE
>
> Full disclosure: I'm color deficient. The final result from what
> comes out of this mix could be undesirable (from a color
> theory standpoint) in our current ecosystem. I recommend
> focusing less on how poor combinations that include colors
> you cannot control might be today, and consider what
> a system might look like where that influence is more
> appropriate and helps produce quality results in the
> near future.

This means that we can support the idea of light mode without
curating much at all. Remember, this is expecting that the user's
settings are accurately conveying what the user wants. This isn't
always the case as mentioned earlier. It's possible they want a
dark mode for a particular part of the experience apart from
other experiences within their interface. This means that we
cannot fully avoid the concept of `data-mode="light"` being

explicitly set by the application, overriding the user setting. We'll need to continue to curate this, and it could just mean that the system is influencing these modes less or not at all:

```
[data-mode~="promo"] {
  --🔒action_primary_backgroundColor: color-mix(
    in oklch,
    oklch(from purple none c h), /* no influence
below, all purple */
    oklch(from ButtonFace l c none) 0% /* no
influence, system color not used */
  );
}
```

The reality is: what we're exploring here is only scratching the surface of what we might consider when allowing the user's preferences to be involved in the final presentation of the experience they are ultimately using. We'll need more support to allow the user's settings and the brand to coexist in a reasonable manner.

User Settings

Even with the inclusion of influence and using the parts of color instead of the complete color, we still lack confidence in the final presentation. This means that folks looking for fine control over the final value of their primary button may be disappointed. Remember, the point here is extreme personalization. A user may curate the value of `ButtonFace` in their system to be wildly unusable to most folks. If that user was expecting to see that curated value in all of their experiences, the brand of any application would cause these to have varying degrees of success. In the majority of cases we would expect the brand influence to be high. This causes whatever the setting used for the system color to be hardly contributing to the final presentation.

A user curating each color specifically for their UI elements is unlikely. In legacy operating systems, user preferences would expose the ability to choose the color of UI elements and areas.

A person could choose a specific color for buttons, text, menus, and other parts of the interface. In more modern operating systems, this has been reduced to choosing a theme or accent color. This is most likely to discourage bad behavior that might make the interface accidentally unusable. Some operating systems discourage this level of customization entirely in favor of their own branded color decisions.

Additionally, the system colors that are currently recognized by applications do not fully cover our current categories and properties of intents. The concept of `ButtonFace` is directly associated with the default `<button/>` element in HTML, which doesn't inherently possess the concept of priority that we discussed earlier. For this system to have the proper coverage, the collection of system colors should be expanded to cover all intents as we've explored.

Perhaps we could support priority as a factor of influence. The higher the priority, the more influence applied to the calculation. This would cause primary buttons to have more branded values than less important buttons. This makes sense, as areas where the brand wants the interface to be prioritized should also be heavily influenced by the brand. Areas that have less priority don't need as much brand influence and could look closer to the standard system colors.

Learning Preferences

For the sake of our exploration, let's lean into the opposite extreme. Let's consider that a user can directly affect how `ButtonFace` and other system colors are finally set within their operating system. This could happen actively, where the user chooses each color meant for each UI element property. This could also happen passively, by a system of learning what the user shows preference for through their choices in interacting with various UI elements over time.

The passive approach seems more appealing at first, as the user does not need to actively curate the colors for their experience.

In theory, this would allow the user to traverse interfaces normally, without curation side quests that are inappropriate to their daily tasks. So the question is: how could we teach the system what colors are most appropriate for a user and avoid being distracted by curation activities?

I believe personalization should not come into effect until we have sufficient data that can better understand this particular user. In other words, we could provide a default `ButtonFace` color in the operating system while it learns which colors are more engaging for buttons. However, unlike the Nest thermostat, the user could affect the setting through the passive behavior of navigating applications, as opposed to consciously adjusting the appearance on every interaction.

Unfortunately, the passive approach has immediate problems. Imagine that this particular user just so happens to only use applications where the button color is blue. This would heavily skew the results toward the blue color such that later influence would assume the user prefers blue. In reality, the blue was an artifact of the choice of product, not the preference of color. The other extreme is where the collection of used applications results in a large spectrum of colors that could produce no useful data. The context tied to the button is heavily influencing the user's choice to interact with it. We would need to separate the task from the purpose of learning preference. This means it is inappropriate to learn within the context of application use; there are too many other factors that would be influencing this one data point.

This further solidifies that the most appropriate way for a system to learn what a user prefers is by deliberately asking. Luckily, unlike the task of understanding purpose in design, we don't need to understand why a user prefers one color over another. The preference is appropriate enough to influence the final result. This is the same as any user choosing dark mode: an organization should justify the creation of dark mode with at least one reason tied to the business. If enough users want dark mode, it could be enough of a business reason to point resources at creating one, regardless of what the user's reasons are.

Note that this is different from usability testing. We're not asking if the experience solves a user problem, we are truly asking if the user prefers one presentation over another from a visual standpoint. We might imagine some opt-in experience that introduces the user to a choice between two expressions of the same application structure. The onboarding would explain that the data is used to learn of the user's visual preferences and the results will be used to influence interfaces over time. The more the system understands what the user prefers, the more personalized the expressions of interfaces could be for this user.

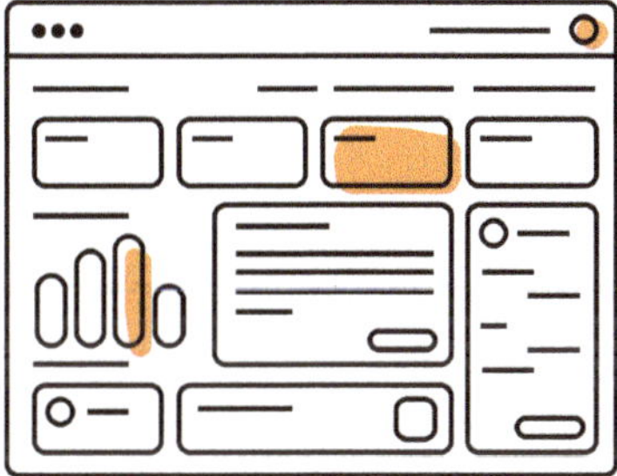

Would you prefer to see all the data at once,
or only the data that is most important

We must also recognize there are outside factors that can influence decisions, such as the time of day. A user might prefer a muted experience in the evening but has only interacted with the learning system during the morning hours. We would want some method of capturing this information without bothering the user when the environment changes. This is the ultimate challenge in learning human behavior; the user's environment should also be under consideration as their choices can be affected by it. Such a considerate system is out of scope for us at the moment—especially as operating systems would need to improve to harness this data. However, keeping in mind that there can be multiple data points contributing to the final decision of preference, we can be prepared to provide a more fine-tuned experience to each user over time.

For now, we'll assume that a user isn't drastically changing the expected colors assigned to `ButtonFace`. We can expect the values provided by the operating system to be reasonable based on other personalization factors, such as light or dark mode.

Unequal Responsibilities

On some websites, there is a custom setting that allows the user to choose light, dark, or the mode provided by the system. In my initial explorations of providing such control, I considered only providing the system mode or the opposite of the system mode. This means that if the system was set to dark, the page would be presented in dark mode. If you use the switch to engage the system's opposite mode, it would present the light mode. The feedback that I received regarding this was that some users prefer certain types of content to be tied to a specific mode. For example, some folks prefer long-form articles to be presented only in light mode. If the system were to dynamically set the mode depending on the time of day, the opposite mode would also switch. These users would prefer to "pin" the selected mode for these sites, effectively overriding whatever the system is suggesting no matter what other conditions were met.

This is an example that challenges our predetermined understanding of the separation between expressions that are operating system responsibilities and ones that are organization responsibilities. In this case, the user is explicitly choosing a different mode based on the content provided. In a future system, this might be a learned data point—but at the moment, this causes the organization to have more responsibility. It can not only provide expressions that enhance the brand and its messaging, but also lower-level modes to support inclusive experiences. In other words: to be truly inclusive, we would need modes that support anything the user needed. Supporting light and dark mode isn't enough. We'd also need to know when to appropriately deliver light or dark mode.

We can see other similar support with content considerations, such as changing the currency when the user purposefully sets their locale within an international shopping website. Settings that a site could read from the operating system might not be specific enough to the current user. We could also see this with guest accounts, though we would expect those to omit any personalization due to the lack of data.

Of course, providing inaccessible colors or font scales that don't convey hierarchy will have detrimental side effects on the experiences relying on these tokens. There is a responsibility held by the person curating these tokens to have values that are made in good faith to our user experience. This has been the case for folks who have been responsible for creating light and dark mode before. Now, we're considering a lot more factors, some of which could be wholly outside of our control for a highly inclusive result. The role of a design system maintainer is to educate the organization on this impact so they can make decisions that influence the user appropriately. The brand's choices could introduce unintended friction or thoughtful delight. All with a single setting.

On the experience design side, there always has been work to do. It's been challenging enough to educate product designers on the idea that their presentation is not necessarily the same on all devices. Taking the user's settings into account is a relatively recent addition to the world of design. We are visual creatures, after all. We'll need to transform all of the theory and possibilities of what could happen, into systems that demonstrate what does happen over time. We'll need tools for designers that show how inclusive design will affect their work. We'll need environments for them to work in that shift the responsibility of how the UI looks away from the core value proposition of the user experience design role. I have

confidence there are better solutions when user problems
are well understood and become the central focus over
aesthetic noise.

It's time for a shift in responsibilities to happen in the practice of
user experience design. The shift should support inclusivity and
accessibility by being influenced from user settings. In the next
chapter, we'll explore what that might look like and how the roles
of our industry might change in the future.

Chapter Summary

While current operating systems and browsers may not yet fully
support the level of personalization needed for truly inclusive
design, we can begin to imagine new approaches to this
challenge. Rather than waiting for perfect system-level
solutions, what if we could develop intelligent systems that learn
and adapt to user needs?

Key Concepts

- Inclusive design requires personalization
- System and brand influences must be balanced
- Privacy concerns affect implementation
- Design roles need to evolve

Next Steps

- Develop better preference learning tools
- Maintain privacy-preserving mechanisms
- Evolve design system roles
- Adopt better design tools

Endnotes

1. *https://resources.arc.net/hc/en-us/articles/19212718608151-Boosts-Customize-Any-Website*

2. *https://en.wikipedia.org/wiki/Nest_Thermostat*

3. *https://developer.mozilla.org/en-US/docs/Web/CSS/system-color*

4. *https://developer.mozilla.org/en-US/docs/Web/CSS/color_value/color-mix*

5. *https://developer.mozilla.org/en-US/docs/Web/CSS/CSS_colors/Relative_colors*

Expression Models

In the previous chapter, we introduced the idea of inclusive design. We also experimented with the possibility of using user data to influence expressions instead of deliberately curating each individual intent in any given mode. It would seem that this has early benefits, because we can offload some of the presentational influence onto the user's device settings, and merely provide a variable amount of brand influence to maintain a company's look and feel.

The reality is: so far, operating systems take a long time to expose accessible data and interfaces to hook into for our purposes. Privacy is a significant topic in this consideration as well, since the data could be used for fingerprinting and other purposes not yet explored. We should be careful when serving hyper-personalized experiences, especially in the case where a user needs support because a design has missed the mark. It can be frustrating when all of the tutorials look different from your current expression.

However, what if we made it easier for users to make influential changes themselves within our UX ecosystem? Is it really so hard to paint with all the colors of the UI?

Algorithmic Curation

What I'm about to propose doesn't exist at the time of this writing, but with the current investments and adoption we've seen in artificial intelligence, the idea could be within our grasp.

You might be familiar with the concept of a "blend mode", which is a method of compositing layers of pixels in photo editing programs. Some common blend modes are "multiply" or "screen", which take the color values of a bottom layer and produce new values through some algorithm, altering the presentation of the image. Each blend mode has a specific algorithm that alters the colors in a repeatable way and you can continue to add additional blend modes to further alter the final result.

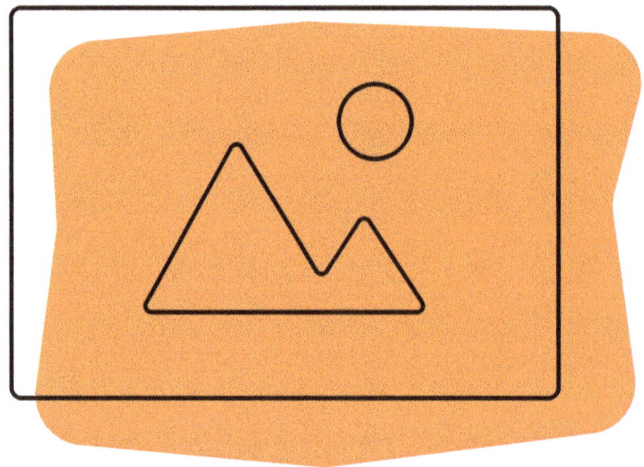

A blend mode layer applied to an image,
altering the final visual presentation

You might have already connected some dots with our current approach of Mise en Mode and the concept of the blend mode. Keeping this concept in mind: instead of curating a specific value for an intent within a mode, what if we curated *algorithms* that were meant to represent the mode that resolved to an appropriate value? For example, what if the critical mode sent base values through an algorithm that was fundamentally a red filter? We've learned earlier that selecting red might not be the culturally correct value for this, so the algorithm would need to be smarter and more aware. However, the general concept is the important point. In the same way, we expect our concept of modes to compound with influences in brand, user preference, and the rest. We can add layers of blend modes that compound algorithms to create a new, final presentation of a base composition.

I'll admit that I wouldn't begin to know what such an algorithm might look like or what form it might take. Realistically, a human-written algorithm would most likely introduce a bias or neglect highly specific nuances which would defeat the purpose of the work. I believe the creation of such an algorithm should be a passive exercise, not an active one. A system shouldn't be explicitly programmed to know how the concept of critical should be expressed; it should learn.

Large Expression Models

You may have heard of the term Large Language Model (LLM) in the subject of artificial intelligence. The LLM is meant to hold an understanding of how language works and have the ability to compose combinations of language that appear to be intelligent. A rudimentary example is entering the phrase "the quick brown fox" into the model. The system will look through its enormous corpus of language to find the most likely candidates that would follow that phrase, returning the one with the highest confidence. Poor models may not complete the phrase or return something nonsensical but more robust models should return "jumps over the lazy dog." How robust the model is depends on the data provided and how well-connected the data is.

Now imagine a Large Expression Model (LEM), that would work similarly to an LLM but with an understanding of how expressions are composed. We could give it some parameters such as "Visa", "Dark", "Critical", and "Chinese", then produce all the values for our intents based on the learned connections and confidence. Instead of curating each value throughout all permutations that could exist, we train the LEM on what it means to express the brand "Visa" and then provide a temperature dial to adjust how much influence that expression might have on an experience. We can do this for any parameter —learning what any particular expression might mean, and then serving a composite mode to the user made just for them.

In this world, people aren't directly responsible for the curation of values to intents, but instead responsible for providing accurate data to teach the LEM how "Visa" is meant to be presented. Certainly, this requires the existence of the "Visa" treatment to exist, but a design system meant to support an organization should not exist without that established brand expression. We should have had a rough agreement regarding the brand earlier on to consider the possibility of a design system at all. Just like how our experiences inform intents, the current presentation of our brand can inform its future through an intelligent system of learning.

A New Role for Design

What I'm proposing here is highly controversial. Earlier I suggested a separation of responsibility between UX and UI design. In this new lens, the UI design part as the role of choosing the values is less about decisions and more about influence. The person who used to curate values is now training LEMs to produce customized results. To be clear, it means that this person no longer assigns the color blue to `$action_primary_backgroundColor`. They provide enough data to the model so the system might learn that blue should most often be assigned to `$action_primary_backgroundColor`, but perhaps not always. When it should be assigned would depend on many factors learned by the LEM.

The role being described isn't anything close to a traditional design role; it is much closer to a data scientist role. This shouldn't deter the possibility of this new role. We often forget about other roles within the experience design practice that aren't primarily visual, such as research, content, and information architecture. A new role called "LUI (Learned UI) analyst" could appear, which expects skills in user interface design concepts along with data analysis techniques. Perhaps you could think of it as the expertise of A/B testing with multiple alphabets of letters simultaneously. Justifying such a role is a matter of selling the benefit of learning what expressions we might provide to motivate users. We would positively grow the company's public perception with verifiable data related to design. Marketing analysts research how to sell the business, and LUI analysts research how to enhance experiences effectively to capture the widest audience of customers possible.

It's important to mention that this isn't replacing every UI designer's role in this new world. Remember how there must be an initial presentation before a system can be grown. The role of traditional UI designers would be to design these initial treatments that will inform the models in the future. You can think of this as the ideal universal presentation. This may be similar to how a UX designer may publish an ideal state of the experience with few variations for the sake of launching a concept quickly. The final implementation of an experience would need to be interpreted by an experienced engineer with several development techniques in mind. The technology is filling in the gaps that could be too numerous to prescriptively fill by one designer in a restrictive design application.

Certainly, it might make sense for this new analyst role to grow from the traditional UI designer role. Thinking about the lifecycle of user interface design above, we expect to start with an initial visual treatment as the baseline. That baseline would be the foundation for changes discovered from the learned system over time. This means that the baseline design work is less than before, while the analysis of future design informs the product over time with changes in compounding factors. Realistically, how often do we create a component never built in any system before? Not often. What happens more often is a desire for new expressions of existing components and patterns. As we covered in earlier chapters, we have infinite possibilities in this space, as long as we aren't clever in our core functionality. Meeting user expectations remains an essential foundation of our practice.

Chapter Summary

We explored how the future of design could move beyond the manual curation of interface values and toward systems that generate expressions through algorithms and learned models, questioning whether designers should continue prescribing every detail or instead guide adaptive systems that respond to brand, user, and contextual inputs.

Using the metaphor of blend modes, the idea of Large Expression Models (LEMs) is introduced. Intelligent systems that, much like language models, could learn what it means to express a brand and produce contextually appropriate results. This shift would redefine the role of the designer: rather than choosing exact values, we would train and influence models, potentially evolving into a new kind of role called the LUI analyst, which blends design intuition with data science.

While traditional design work still provides the necessary foundation, long-term opportunity lies in preparing for a world where expressions are dynamically learned, not statically assigned.

Key Concepts

- Algorithmic design decisions
- Blend mode model of expressions
- Large Expression Models (LEMs)
- Evolution of design roles

Next Steps

- Shift from manual to algorithmic curation
- New role of LUI analyst
- Data-driven design decisions
- Partnership between AI insight and human design

Future of Expression

Throughout this book, we've explored a fundamental shift in how design systems handle visual expression. From static tokens to dynamic modes, from manual curation to intelligent adaptation. This journey has challenged long-held assumptions about control, consistency, and the very purpose of design systems themselves.

From Control to Influence

Traditional design systems were built on a premise of control. They prescribed exact colors, specific spacing values, and rigid component implementations. The goal was consistency through constraint. But this approach created an unintended consequence: it limited the system's ability to adapt to the diverse needs of users and contexts.

We've discovered a different path. By separating structure from expression, we enable systems that are both consistent *and* flexible. The breakthrough insight is deceptively simple: **design systems shouldn't control expression, they should influence it through well-structured intentions.**

Intents form the foundation of this approach. Rather than defining that a primary button must use a specific hex value, we declare that it should express primary importance. This requires a clear separation of purpose from presentation, a structured understanding of priority levels that convey meaning rather than just visual weight, and a scalable token architecture that grows without accumulating complexity. The result is a mode-based expression system that adapts to context while maintaining semantic coherence.

This shift from "the button must be teal" to "`$action_primary_backgroundColor` expresses the highest priority of interaction for a user" unlocks extraordinary flexibility while maintaining the consistency that design systems promise. It's not about abandoning standards; it's about making those standards smart enough to adapt.

The Transformation of Modes

Mise en Mode introduces a fundamental change to how design systems handle variation. The approach moves us from tokens as values to dynamic expressions. Multiple influences can combine seamlessly. Imagine a warm, high-contrast dark mode that maintains brand identity while serving a user's specific needs. In a near future, contextual adaptation could happen automatically, without manual intervention. Brand identity remains consistent, even as expression varies wildly across different modes and user preferences.

The impact on real-world problems is profound. Token bloat decreases dramatically as you no longer need separate tokens for every variation. The maintenance burden drops significantly when you're managing intentions rather than hundreds of hardcoded values. Customization scales efficiently because new modes can be added without exponentially increasing system complexity. Perhaps most importantly, inclusive design becomes a natural improvement. Features like high-contrast or reduced-motion aren't special accommodations—they're just another mode of the same underlying system.

Where traditional systems might manage hundreds of token variants manually, each requiring individual updates and careful coordination, mode-based systems can support any combination of influences to produce an appropriate expression for the moment. The system becomes generative rather than prescriptive.

The Intelligence Revolution

As we look toward the horizon, we hope artificial intelligence emerges not as a replacement for design thinking, but as an evolution of how design systems operate. Large Expression Models (LEMs) represent a paradigm shift as significant as the move from manual layout to responsive design.

LEMs enable dynamic generation of expressions rather than manual curation. Instead of designers creating every possible variation, algorithmic decision-making determines style choices based on context, user preferences, and brand requirements. These systems learn from user preferences to improve over time, and they automatically balance brand requirements with individual user needs in ways that would be impossible to manage manually.

This suggests a future where expression is generated in real-time rather than predetermined during the design phase. Personalization happens automatically without requiring designer intervention for every edge case. Brand identity becomes a directional influence—a set of constraints and preferences—rather than rigid, unchangeable rules. Design systems transform from static libraries into living, learning platforms that evolve with their users.

The implications are staggering. Instead of designing for the "average user"—a statistical fiction that represents no one—we

can create systems that adapt to each individual's needs,
preferences, and contexts while maintaining brand coherence. A
user with low vision gets the contrast they need. Someone
sensitive to motion gets gentler animations. A user
working in bright sunlight gets higher luminance. All of this
happens automatically, all within the bounds of the brand's
expression language.

Evolution of Design Practice

This technological transformation demands an evolution in how
we think about design roles and processes. The skills that made
someone an excellent UI designer in the past decade may not be
the skills most valuable in the coming years.

UI designers are evolving into expression architects who define
the boundaries and influences of generated design rather than
every pixel. They set the parameters within which the system
can operate, establish the principles that guide algorithmic
decisions, and ensure that generated outputs serve user needs
while expressing brand identity. Design system maintainers
become facilitators who nurture the system's learning and
adaptation rather than gatekeepers who approve every change.
They monitor how the system performs, identify opportunities
for improvement, and guide its evolution over time.

New specialists will emerge, like LUI (Learned User Interface)
analysts who interpret and guide expressive decisions. These
practitioners bridge the gap between design intuition and
machine learning, helping systems understand what users do
and why they do it. The focus across all these roles shifts from
pixel-perfect control to intentional influence and strategic
direction. Designers stop being the bottleneck through which all
changes must flow and become the architects of systems that
make good decisions independently.

Let's be clear: this isn't about replacing designers. It's about
unlocking new creative possibilities. Everything that makes you
valuable as a designer—your empathy for users, your eye for

detail, your ability to wrangle complexity into clarity—all of that matters more than ever. What's different is the scale at which you work. You're not just designing screens anymore; you're designing the intelligence that designs screens. Your skills don't diminish; they multiply across every user, every context, every moment your system touches someone's life.

Principles for the Future

While some concepts in this book may feel futuristic, the foundations can and should be implemented today. Intents provide immediate benefits in maintainability and clarity. Modes scale better than traditional approaches from day one. Infrastructure investments pay dividends long before AI enters the picture.

The evolution of design roles doesn't require waiting for LEMs to mature. It begins the moment you start thinking about influence rather than control, about intention rather than prescription. Every decision to document the "why" behind a design choice, every token renamed to express intent rather than appearance, every mode added to support user needs. These are steps on the path toward intelligent, adaptive design systems.

As design systems continue evolving, hold fast to these core principles. Support human expression. Technology should amplify human creativity and meet human needs, not constrain them to fit technical limitations. Enable inclusive experiences where every user deserves interfaces that work for their specific needs and preferences, not just those of the majority. Adapt to user needs, recognizing that rigidity is a bug rather than a feature, and systems should flex to accommodate the diversity of human experience.

Maintain brand identity even as you embrace flexibility. Flexibility doesn't mean chaos, and brand coherence can coexist with personalization when the system is structured thoughtfully. Scale efficiently so that as your organization and user base grow, your design system becomes more powerful rather than more burdensome, enabling rather than constraining the teams that depend on it.

Final Thoughts

The journey from static design systems to intelligent, adaptive platforms is just beginning. The challenges ahead are real: technical complexity that pushes the boundaries of current tooling, organizational change that requires new ways of working, ethical considerations about algorithmic design and who gets to decide what's appropriate, questions about the role of human judgment in increasingly automated systems.

But the opportunity is extraordinary. We have the chance to build design systems that truly serve everyone. Designs that adapt to individual needs without sacrificing coherence, that scale efficiently, and that support inclusive design. We can create experiences that feel personally crafted for each user while maintaining the efficiency and consistency that design systems promise.

Build for influence, not control. Design for adaptation, not prescription. Create systems that evolve with both technology and human needs. The future of expression is dynamic, intelligent, and deeply human. Let's build it together.